Supported by the EUROPEAN COMMISSION

Directorate General XII
Science, Research and Development

EUR 16823

LEGAL NOTICE

The book is based on work done in a number of R&D projects in the framework of an ongoing programme of the European Commission. This programme, called JOULE, supports research and development into renewable energy technologies. It is managed by the Directorate General XII for Science, Research and Development. The report, as published in this book, is supported by the Commission under EC-Contract JOU2-CT92-0148.

Cataloguing data can be found at the end of this publication.

Springer
Berlin
Heidelberg
New York
Barcelona
Budapest
Hong Kong
London
Milan
Paris
Santa Clara
Singapore
Tokyo

S. Wagner, R. Bareiß, G. Guidati

Wind Turbine Noise

With 88 Figures

Prof. Dr.-Ing. Siegfried Wagner
Dipl.-Ing. Rainer Bareiß
Dipl.-Ing. Gianfranco Guidati

Universität Stuttgart
Institut für Aerodynamik und Gasdynamik
Pfaffenwaldring 21
D-70550 Stuttgart/Germany

European Commission (DGXII), EUR 16823

ISBN 3-540-60592-4 Springer-Verlag Berlin Heidelberg NewYork
ISBN 0-387-60592-4 Springer-Verlag NewYork Berlin Heidelberg

Die Deutsche Bibliothek - CIP-Einheitsaufnahme
Wagner, Siegfried: Wind turbine noise / S. Wagner; R. Bareiss; G. Guidati. -
Berlin; Heidelberg; New York; Barcelona; Budapest; Hong Kong; London; Milan;
Paris; Santa Clara; Singapore; Tokyo: Springer, 1996
ISBN 3-540-60592-4 (Berlin ...)
ISBN 0-387-60592-4 (New York ...)
NE: Bareiss, Rainer; Guidati, Gianfranco

Printed in Germany

Typesetting: Camera ready by author
SPIN: 10522931 61/3020 - 5 4 3 2 1 0 - Printed on acid-free paper

Preface

Within the last five years an enormous number of wind turbines have been installed in Europe, bringing wind energy into public awareness. However, its further development is restricted mainly by visual impact and noise.

The noise regulations of various countries urge turbine manufacturers to reduce the noise emission of their turbines. Therefore, the Commission of the European Union addressed the problem of wind turbine noise within five research projects which were funded within the framework of the JOULE program (JOULE - Joint Opportunities for Unconventional and Long term Energy Options), see Table I. These projects covered different aspects of wind turbine noise.

The need to disseminate the results of the noise projects to a large number of people, namely researchers, manufacturers, and planners, motivated the publication of this book on *wind turbine noise*. It is based on considerable research and evaluation of literature at the Institute of Aerodynamics and Gasdynamics (IAG) as well as on the work performed within the projects and presents the most relevant results. A complete description of the work can be found in the final reports [24], [120], [131], [203].

Table I: Projects on wind turbine noise funded by the European Commission.

Title	*Contract number*	*Coordinator*
Aerodynamic noise from wind turbines	JOUR-CT90-0107	DELTA, Dk
Noise from wind turbines; variability due to manufacturing and propagation	JOU2-CT92-0124	ECN, NL
Development of an aeroacoustic tool for noise prediction of wind turbines	JOU2-CT92-0148	IAG, D
Investigation of blade tip modifications for acoustic noise reduction and rotor performance improvement	JOU2-CT92-0205	ICA, D
Aerodynamic noise from wind turbines and rotor blade modifications	JOU2-CT92-0233	DEWI, D

The work performed in the above mentioned noise projects was co-financed by the European Commission DG XII. Other parts of the budget have been financed by *Centre for Renewable Energy Sources, Danish Ministry of Energy (EFP), DELTA Acoustics & Vibration, Deutsche Forschungsanstalt für Luft- und Raumfahrt (DLR), Deutsches Windenergie-Institut GmbH, dkTeknik, Flow Solutions Ltd., National Engineering Laboratory (NEL), National Aerospace Laboratory (NLR), Netherlands Energy Research Foundation (ECN).*

The authors are indebted to all participants of the JOULE II projects for their contributions and fruitful discussions. Special thanks goes to: *Wolfgang Palz* and *Komninos Diamantaras* of the European Commission for their valuable support of the aeroacoustic projects and funding of the projects which enabled the publication of this book, *Ton Dassen* for reviewing the book and writing several sections about noise measurements and noise reduction, *Jan Bruggeman* for reviewing the book and giving us very helpful comments and recommendations at an early stage of our work on aeroacoustics, *Rene Parchen* for helpful discussion about the problems of aeroacoustics, *Jørgen Jakobsen* and *Jørgen Kragh* for their support and contributions to the chapter about noise propagation, *Nico van der Borg* for writing sections on acoustic outdoor measurements, *Kurt Braun* and his group for writing sections on flow visualization, *Erdmuthe Raufelder* and *Springer-Verlag* for valuable support during the preparation of the manuscript.

On behalf of all participants of the various noise projects, we wish to thank the coordinators *Kurt Braun* (ICA project), *Helmut Klug* (DEWI project), *Nico van der Borg* (ECN project), *Bent Anderson* (DELTA project), and *Frank Hagg* (TWIN project). Our thanks goes as well to *Spyros Voutsinas* and *Serge Huberson,* who worked together with the authors in the IAG project. We would like to thank *Doris Hug, Jutta Guidati, Vincenzo Guidati, Otto W. Bareiß* and all those not explicitly mentioned who have contributed with support, comments, and recommendations.

Furthermore, we wish to thank the following organizations for supplying us with valuable information: *Forschungsstelle für Energiewirtschaft München, Preussen Elektra AG, Deutsche Verbundgesellschaft, Umweltbundesamt Berlin, Deutsches Windenergie-Institut GmbH.* Finally, we gratefully acknowledge the assistance of *Ursula Henn, Jagjit Kaur,* and *Vernon Robinson* who directly contributed to the preparation of the manuscript. Last but not least, we thank our wives for their support and patience.

Siegfried Wagner
Rainer Bareiß
Gianfranco Guidati

Stuttgart, February 1996

Summary

The purpose of this book is to give an overview of state-of-the-art knowledge in the field of wind turbine noise. Chapter 1 starts with a survey of the current situation of wind energy in Europe. The advantages of wind energy are discussed together with the hindrances to promote this technology. Chapter 2 introduces general definitions that are used for describing noise and its effect on people. Chapter 3 contains an introduction to the theory of aerodynamic noise generation. Special emphasis is put on the topics which are relevant for wind turbine noise, mainly noise from turbulence (Lighthill's acoustic analogy) and its interaction with solid surfaces.

In Chapter 4, the mechanisms that create noise on wind turbine blades are explained. The importance of the different mechanisms is discussed and the main influencing factors are outlined. Chapter 5 gives an overview of the different approaches to noise prediction. This includes simple rules of thumb which can be used for an overall estimate of the noise level and more sophisticated models which consider details of the flow around a blade section. It becomes clear that a prediction model which can be used to design low-noise airfoils is not available at the moment. However, follow-up work in the framework of JOULE III projects (see Chapter 9) is directed towards this aim.

The possibilities and limitations of noise propagation methods are described in Chapter 6. Recent advances in this field are included as well. Chapter 7 gives an introduction to the techniques for noise measurements on wind turbines and in the wind tunnel. Furthermore, the possibilities of flow visualization on wind turbine blades are tackled.

Chapter 8 discusses the important topic of noise reduction. It starts with rather simple means like reduction of tip speed or angle of attack which, however, have a negative impact on the produced power. It continues with modifications at the blades, namely the trailing edges and the tips. This includes serrated and beveled trailing edges and a variety of different tip planform shapes. The effectiveness of the different means for noise reduction is discussed based in part on theoretical findings and mainly on experimental work which has been performed within the EU-funded projects (see Preface).

Chapter 8 is thought to be particularly valuable for blade and turbine manufacturers.

The book closes with some recommendations for future research. Interesting topics are an improvement of the understanding and modeling of noise generating mechanisms and the development of efficient means for noise reduction. This research will include both theoretical and experimental work.

Contents

Nomenclature

Symbol	*Unit*	*Description*
a_m^F	N	Fourier coefficients of *m*th axial force harmonic
a_m^M	Nm	Fourier coefficients of *m*th torque harmonic
A	Pa	Amplitude
A_1, A_2	Pa	Complex amplitudes in general
A_b	m^2	Blade area
A_R	m^2	Rotor area
c	m/s	Wave speed
c_0	m/s	Acoustic wave speed, speed of sound
C_r	m	Chord at hub, root
C_{ref}	m	Reference chord length at 30 % blade span $(R-R_h)$
C_t	m	Chord length at tip
C_T	-	Axial force coefficient, thrust coefficient
C_i	dB	Constants for rule of thumb noise prediction formulas
d	m	Distance from source to shadow point
d_i	m	Distance from receiver to the point where the sound ray touches the ground
d_s	m	Distance from source to the point where the sound ray touches the ground
D	m	Rotor diameter
D_{BM}	dB	Damping due to ground effect
D_D	dB	Damping due to vegetation
D_e	dB	Damping due to noise screens
D_G	dB	Damping due to obstacles and buildings
DI	dB	Directivity factor
D_L	dB	Damping due to air absorption
D_s	dB	Geometrical divergence
$\overline{D}_1, \overline{D}_2, \overline{D}_h$	-	Directivity functions
f	Hz	Frequency
f	-	Function for the description of surfaces (see Section 3.6.1)
$\vec{f}$	N	Concentrated point force
f_B	Hz	Blade passing frequency
f_c	Hz	Critical frequency separating high- and low-frequency region for inflow turbulence noise
$\vec{f}_i$	N	Aerodynamic forces at source co-ordinates *i*
f_{max}	Hz	Maximum frequency
f_n	Hz	Frequency corresponding to the *n*th harmonic
f_{peak}	Hz	Peak frequency
f_R	Hz	Rotor frequency
$\vec{F}$	N/m^3	Force per unit volume
g	-	Function in general
$\tilde{g}$	-	Fourier transform of function *g*
G	-	Green's function

G_1	dB	Empirical scaling function representing the influence of Strouhal number on LBL/VS noise
G_2	dB	Empirical scaling functions representing the influence of Reynolds number on LBL/VS noise
G_3	dB	Empirical scaling functions representing the influence of angle of attack on LBL/VS noise
G_4	dB	Empirical scaling functions determining the peak level for BTE noise
G_5	dB	Empirical scaling functions determining the shape of the spectrum for BTE noise
G_6	dB	Empirical scaling functions determining the shape of the spectrum for TBL/TE noise in Lowson's model
G_A, G_B	-	Strouhal number based scaling functions
h	m	Hub height
h_s	m	Source height
h'_s	m	Modified source height
h_i	m	Receiver height
h'_i	m	Modified receiver height
$H(x)$	-	Heaviside function (see Section 3.2.3)
i	-	Source number, counter, index, imaginary number, $i = \sqrt{-1}$
I	W/m^2	Acoustic intensity
$\vec{I}$	W/m^2	Vector of acoustic intensity
I_{ref}	W/m^2	Reference acoustic intensity, 10^{-12} W/m^2
j	-	Index, counter
J_x	-	Bessel function of first kind and order x
k	1/m	Wave number, $k = 2\pi/\lambda$
k_i	1/m	ith component of the wave vector
k_0	1/m	Acoustic wave number
$\vec{k}$	1/m	Wave vector, (see Section 3.2.1)
$\widehat{k}$	-	Normalized wave number in Lowson's IT noise model, see [3]
K_0	dB	Reflection factor
$K_1(f)$	dB	Scaling function for IT noise, see Fig. 5.7
K_2	dB	Empirical constant = 3.5
$K_3(f)$, $K_4(f)$	dB	Scaling functions for TBL/TE noise, see Fig. 5.7
K_5, K_6, ΔK_6	dB	Empirical functions for TBLTE in BPM model
K_{lfc}	-	Low-frequency correction factor in Lowson's IT noise model
l	m	Length in general, typical dimension
l_{IT}	m	Overall scale of atmospheric turbulence in Lowson's IT model
l_{tv}	m	Extension of tip vortex
L_{dn}	dB	Energy averaged sound pressure level
$L_{eq,T}$	dB	Equivalent sound pressure level
L_p	dB	Sound pressure level
L_{pA}	dB(A)	A-weighted sound pressure level

$L_{p,\mathrm{BTE}}$	dB	Sound pressure level of BTE noise
$L_{p,\mathrm{IT}}$	dB	Sound pressure level of IT noise
$L^{\mathrm{H}}_{p,\mathrm{IT}}$	dB	High-frequency sound pressure level of IT noise
$L_{p,\mathrm{LBLVS}}$	dB	Sound pressure level of LBL/VS noise
$L_{p,\max}$	dB	Maximum sound pressure level in the spectrum
L_{pn}	dB	Sound pressure level of *n*th noise harmonic
$L_{p,\mathrm{TBLTE}}$	dB	Sound pressure level of TBL/TE noise
L_W	dB	Sound power level
L_{WA}	dB(A)	A-weighted sound power level
m	-	Load harmonic for low-frequency-noise calculation
M	-	Mach number, $M = U/c_0$
M_c	-	Convection Mach number = $0.8 \cdot M$
$\vec{M}_i$	-	Local source Mach number = $\frac{1}{c_0} \cdot \frac{\partial \vec{y}_i}{\partial t}$
M_r	-	Relative Mach number = $\vec{r}_i \cdot \vec{M}_i$, $M_r = \frac{\vec{x}-\vec{y}}{r} \frac{1}{c_0} \frac{\partial \vec{y}}{\partial t}$
M_{tv}	m/s	Mach number based on maximum velocity within separated flow region around the trailing edge
n	-	Number of sound pressure harmonic(s)
$\vec{n}$	-	Unit vector, direction in general
n_B	-	Number of blades
n_s	-	Number of acoustic sources on rotor
$\vec{n}_S$	-	Outward normal unit vector of surface S
p, p'	Pa	Acoustic pressure, pressure
$\hat{p}$	Pa	Root mean square of sound pressure
$\hat{p}_{\mathrm{ref}}$	Pa	Root mean square of reference pressure = $2 \cdot 10^{-5}$ Pa
p_0	Pa	Pressure of the undisturbed medium
$p_{\mathrm{lf,i}}(t)$	Pa	Acoustic pressure of far field loading noise at source i
$p_{\mathrm{ln,i}}(t)$	Pa	Acoustic pressure of near field loading noise at source i
p_n	Pa	Sound pressure level of the *n*th noise harmonic
$p_{\mathrm{t,i}}(t)$	Pa	Acoustic pressure of thickness noise at source i
P	W	Acoustic power
P_{ref}	W	Reference acoustic power = 10^{-12} W
P_{WT}	W	Rated power of the wind turbine
q	kg/s	Inflow-rate of matter
Q	kg/(m³s)	Inflow-rate density of matter
r	m	Radius, distance source to observer, $r = \lvert \vec{x} - \vec{y} \rvert$
$\vec{r}$	m	Vector from source to observer, $\vec{r} = \vec{x} - \vec{y}$
$\vec{r}_i$	-	Unit vector pointing from source i to observer = $\frac{\vec{x}-\vec{y}_i}{r}$
R	m	Rotor radius

R	kJ/(kgK)	Specific gas constant of air
R_i	m	Rotor radius at source i
Re	-	Chord-based Reynolds number
Re_0	-	Chord-based reference Reynolds number
R_{ref}	m	Reference radius, 70 %
s	m	Airfoil span
Δs	m	Airfoil segment span
S	m^2	Surface in general
St	-	Strouhal number
St'	-	Strouhal number based on BL thickness at TE
St''	-	Strouhal number based on extension of tip vortex = $fl/U_{\max}$
St'_{peak}	-	Reference Strouhal number based on BL thickness at TE for LBL/VS noise
St'''_{peak}	-	Reference Strouhal number for BTE noise
St_1, St_2	-	Empirical reference Strouhal numbers for TBLTE in BPM model
$St_{\max}$	-	Maximum Strouhal number = 0.1
t	s	Time, observer time
t^*	m	Trailing-edge thickness
t_{ij}	kgm^2/s^2	Concentrated Lighthill-tensor
T	s	Period, inverse of frequency f
T	K	Thermodynamic temperature
T_{ij}	$\text{kg/(ms}^2)$	Lighthill-tensor
u_n	m/s	Component of flow velocity outward normal to S
$\vec{u}$	m/s	Vector of flow velocity
$\vec{u}'$	m/s	Acoustic particle velocity
U	m/s	Free stream velocity, wind speed component
U_c	m/s	Convection velocity
U_{tv}	m/s	Maximum velocity within separated flow region around the trailing edge
V	m^3	Volume
V_0	m^3	Blade volume
V_{rot}	m/s	Velocity due to rotation at r, R_{ref}
V_{Tip}	m/s	Tip speed at rotor blade
V_w	m/s	Wind speed
$\overline{w}^2$	m^2/s^2	Mean square turbulence level
w_r^2	-	Reference turbulence intensity
x	m	Space co-ordinate in general
$\vec{x}$	m	Field point in general, observer position
x_i	m	Space co-ordinate in the ith-direction
$\vec{y}$	m	Source point
$\vec{y}_i$	m	Co-ordinates of acoustic source i
z	m	Height above ground
z_0	m	Roughness height
Z_0	$\text{kg/(m}^2\text{s)}$	Specific acoustic impedance, $Z_0 = \rho_0 c_0$

α	-	Normalized turbulence intensity
β	-	Compressibility factor = $1\text{-}M^2$
γ	°	Observer angle
δ	m	Boundary-layer thickness
$\delta(x)$	1/m	Delta function (see Section 3.2.3)
$\delta(\vec{x})$	$1/m^3$	Three-dimensional delta function (see Section 3.2.3)
δ^*	m	Displacement thickness
δ^*_{avg}	m	Average displacement thickness
δ_{ij}	-	Kronecker-delta, $\delta_{ij} = \begin{cases} 1 & if \quad i = j \\ 0 & if \quad i \neq j \end{cases}$
δ_p^*	m	Displacement thickness at pressure side
δ_s^*	m	Displacement thickness at suction side
$\varepsilon, \varepsilon_1, \varepsilon_2, \varepsilon_3,$	m	Eddy size, small distances
θ	°	Ray arrival angle
κ	-	Adiabatic gas exponent, for air = 1.4
λ	m	Wavelength
μ	kg/(ms)	Viscosity
π	-	The number Pi
ρ	kg/m^3	Fluid density
ρ'	kg/m^3	Acoustic density
ρ_0	kg/m^3	Density of the undisturbed medium
σ	$kg/(m^3s^2)$	Right-hand side of the wave equation
ΣD	dB	Sum of sound reducing effects during noise propagation
τ	s	Source time, retarded time, emission time
$\Delta\tau$	s	Increment of source (retarded) time
ϕ	°	Directivity angle, see [82]
ψ	°	Directivity angle, see [82]
ψ_{TE}	°	Trailing-edge angle of airfoil
ω	Hz	Angular frequency $\omega = 2\pi/T = 2\pi f$
$\vec{\omega}$	1/s	Vector of vorticity, $\vec{\omega} = \nabla \times \vec{u} = rot(\vec{u})$
$\vec{\Omega}$	rad/s	Vector of rotor rotation
$\wp$	-	General symbol including various parameter which influence noise
$\Re$	-	Radius of sound ray
$\mathcal{S}$	-	Sears / airfoil response function

Abbreviations

Symbol	*Description*
BET	Blade-element-momentum theory
BL	Boundary layer
BPM	Brooks, Pope, Marcolini [30]
BTE	Blunt-trailing edge
DFN	Discrete-frequency noise
HAWT	Horizontal-axis-wind turbine
HFN	High-frequency noise
IT	Inflow turbulence
LBL/VS	Laminar-Boundary-Layer-Vortex-Shedding Noise
LFN	Low-frequency noise
SPL	Sound-pressure level
TBL/TE	Turbulent-boundary-layer-trailing-edge interaction
TE	Trailing edge
VAWT	Vertical-axis turbine

In figures and tables the units are given in round brackets, e.g. "convection velocity (m/s)". For non-dimensional quantities only the name is given, e.g. "Strouhal number".

Organization of the Book - Target Groups and Icons

This book is addressed to 3 different target groups: (1) planning companies, authorities, utilities, (2) research institutions, universities and (3) industry: blade/turbine manufacturers. Three different icons are assigned to the headlines of each chapter in order to help the reader to find the relevant topics. The icons appear in two forms: *enabled*, indicating special interest for the marked group, or *disabled*, see Table II.

Table II: Meaning of the icons.

Target group	*Enabled*	*Disabled*
Planning companies, authorities, utilities		
Research institutions, universities		
Industry: blade/turbine manufacturers		

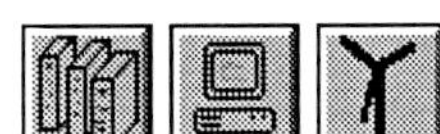

1 Introduction

1.1 Current Situation of Wind Energy and Perspectives

In 1991, the world-wide electricity production amounted to $12030 \cdot 10^9$ kWh, being $20450 \cdot 10^9$ kWh in 2010 as projected by the IEA [180]. The resources of current energy sources are limited, i.e. uranium will be available until 2050–2100, coal until 2110–2190, gas until 2025–2050, and oil until 2020–2030 [180]. Together with the need for reducing air pollution, this leads to a need for increasing contribution of the renewable energy sources.

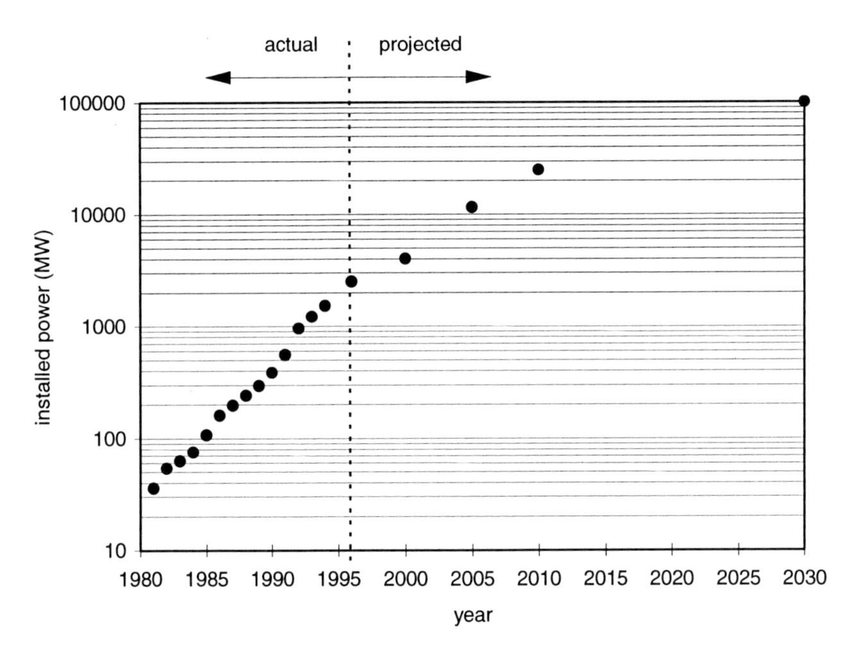

Figure 1.1: Installed wind energy capacity in Europe – actual and projected [36], [70].

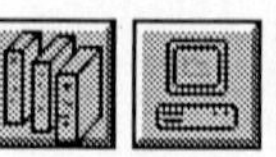

Wind energy, being one of those sources, has made tremendous progress in the last 15 years, especially in the European Union where, according to information of the European Commission (EC - DG XII /F/4 - kd), the installed capacity reached a total amount of 2504 MW (January 31, 1996), compared to 1770 MW in the USA (October 31, 1995) and 4720 MW world-wide (October 31, 1995), see Figure 1.1. Thereby a new industrial sector has developed, creating new jobs, and new market opportunities have been created. Today's wind energy scene, as presented by K. Diamantaras and W. Palz, can be summarized as [47]:

- The wind energy equipment market has stabilized within the last two years to 400 MW per annum providing a stable base for further developments.
- The production cost of commercially available wind turbines at good windy sites is approximately 0.05 ECU/kWh which is pushing wind energy further towards self-supportance.
- Wind energy creates jobs: 4 jobs/$MW_{manufactured}$ and 1 job/$MW_{installed}$ for maintenance.
- The economic potential of wind power increases in line with the new emerging markets and the technological improvements achieved.
- Visual impact, noise problems, and planning issues remain the major obstacles to the further exploitation of wind energy.
- Further R&D and demonstration is required to overcome remaining technological and scientific problems.
- Appropriate technology push and market pull actions are required to maintain the current trends.

This expansion was brought about with the aid of large programs of research, development, and demonstration on the part of the member states and the European Commission. There are a number of research and technological development programs in the field of Non-Nuclear Energy that have been supported by the Commission for more than a decade [9]. In particular the program named JOULE (**J**oint **O**pportunities for **U**nconventional and **L**ong Term **E**nergy Options) was focused on energy-related research areas.

Based on the support of R & D activities and the current wind power technology, the future holds a prospering outlook. Garrad [70] cites the major goal of the European Wind Energy Association, to generate 10 % of Europe's electricity from wind by the year 2030. Intermediate goals are 4000 MW in the year 2000 and about 11.500 MW in the year 2005. The 10 % of Europe's electricity mentioned above is equivalent to 100.000 MW [174].

The estimated cost figures for renewable energy sources by the year 2000 are given in Table 1.1 according to the Renewable Energy Study by the EUREC Agency [174].

Table 1.1: Estimated costs for renewable energy sources by the year 2000 [174].

Renewable energy source	*Estimated costs* (ECU/kWh)
Wind	0.03
Solar Photovoltaics	0.30
Biofuel	0.05

These figures can be compared to approximately 0.04 ECU/kWh required for conventional power sources [174].

With the existing infrastructure, Europe is likely to become a major site for exploitation of wind energy in the near future where along with onshore wind energy resources, offshore is also attracting attention [165]. Many studies on offshore wind turbines were carried out by Denmark, the Netherlands, Sweden, United Kingdom, the International Energy Agency, and the European Union [72]. In 1994, a detailed study within the JOULE program was completed for 11 European Union coastal member countries [165]: The offshore potential ranges from around 500 to 2000 TWh/year for a water depth ranging from 10 to 40 m within a 10 km distance from the shore. The comparison with the annual consumption within the EU of 1845 TWh/year shows the enormous offshore wind potential.

1.2 Advantages of Wind Energy

In addition to the enormous offshore potential, wind energy offers further benefits making it very attractive compared to conventional energy sources:

- *Air pollution* is the major drawback of conventional fossil energy sources. Each kWh of electricity produced by wind energy saves around 1 kg of carbon dioxide. This means that replacing 1 % of the EU's fossil burning electrical generating capacity by wind energy would avoid around 15 million tons of carbon dioxide emission [70]. The same is true for other air pollutants, see Table 1.2.
- Compared to energy harvest factors of coal plants and nuclear power plants being 71.4 and 108 [185] respectively, wind energy shows smaller values, but without using any fossil or nuclear resources during operation. The *energy harvest factor* is defined as the ratio of the energy produced during the lifetime of a wind turbine in relation to the energy used for production, transport, recycling etc. In [103] values, ranging from 38 to 84, are given for a 450 kW turbine, Garrad [70] gives a value of 20, Schaefer [185] values of 10 to 30.

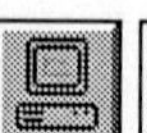

- Furthermore, wind is a clean source of energy which is free from ironising radiation, radioactive waste and does not impose external costs such as costs of air pollution, damage to public health, military protection of fuel supplies, cleaning up after oil spillage etc. External costs for energy produced by coal plants are estimated to be 0.04 ECU/kWh and are not included in current energy prices [70].
- The projected *costs* of wind energy of 0.03 ECU per kWh by 2000 (see Table 1.1) demonstrate the capability of wind energy to compete with conventional energy prices of 0.04 ECU per kWh [174].

Table 1.2: Avoided air emission by using electricity generated by wind energy.

Pollutant	*Amount saved* (kg/kWh) [71]	*Pollution avoided per year replacing 1 % of the EU's fossil burning electrical generating capacity by wind energy* (tons/year) [70]
Carbon dioxide, CO_2	0.750 - 1.250	15000000
Flue dust	0.040 - 0.070	1000000
Sulfur dioxide, SO_2	0.005 - 0.008	20000
Nitrous oxide, NO_x	0.003 - 0.006	40000

1.3 Current Problems of Wind Energy

Noise and visual impact caused by wind turbines give rise to problems concerning public acceptance of wind energy. This is due to the fact that Europe is densely populated in contrast to many countries outside the EU. Compared to the USA with 27 inhabitants per km^2, population density in Europe is with one exception much higher, i.e. D: 223, F: 104, GB: 235, I: 191, E: 79, NL: 358, B: 327, GR: 78, P: 113, Dk: 120, IRL: 50, L: 147, AU: 93, SW: 19 inhabitants per km^2.

This is aggravated by the decreasing number of economic sites in Europe which urges organizations involved in planning to place turbines closer to the populated areas. Furthermore, due to relatively high mean wind speeds, the commercial use of wind energy is especially attractive in coastal areas, which in a lot of instances are places of natural beauty and in some cases wild life reservates.

1.3.1 Wind Turbine Noise

A wind turbine in operation produces noise which may be a cause of annoyance for people who live close to the turbines. Incidents have been reported [124], [196] where noise from wind turbines has been cited as "annoyance" or "nuisance".

The perceived noise at a given observer location depends on the turbine construction, its operation, and situational factors [212] (see Chapters 4, 5, 6), namely:

- the distance between the wind turbine and the living area,
- background noise level at the proposed site of turbine observer location,
- the location of the residence relative to the site with regard to the wind direction,
- other buildings between the wind turbine and the subject house,
- natural barriers such as hills and trees between the subject house and the wind turbine,
- operating conditions of the wind turbine, i.e. wind speed, rotational speed etc.,
- wind turbine components such as tower, tip, and airfoil shape,
- characteristics of the noise source, i.e. tonality, impulsive character.

Psychological factors can have an influence on the acoustical and visual perception of wind turbines and the attitude towards wind energy in general. Some people are more susceptible than others depending on physical health, personality, mood, etc. Psychological research on reactions to environmental noise as reported by Fileds [61] revealed that:

- The awareness of non-noise problems *increases* annoyance.
- Fear of the noise source *increases* annoyance.
- The belief that the noise could be prevented *increases* annoyance.

On the other hand, Fileds states a *decreasing annoyance* to people if they are convinced that operation of the noise producing machine is important.

Lee et al. [138] conducted a study in UK, where 35 % of the respondents felt that a wind farm would spoil the view, 75 % of respondents felt that, such as electricity pylons, wind turbines were ‘just there’, and 90 % of respondents preferred turbines to be painted in neutral colors. Garrad [69] states

> There is a "NIMBY" - **n**ot **i**n **m**y **b**ack **y**ard - attitude over wind energy as over the majority of industrial developments. However, unlike most other developments the public is, given its increasing greenness, basically well disposed to the development of wind energy. From experience in Denmark and the Netherlands the successful development of wind farms seems to depend on providing the public with enough

> information to be able to make an informed decision. The tendency to approach wind farm developments with a "big brother" attitude has been disastrous and must be avoided. It is important to capitalize on people's positive attitudes towards wind energy as opposed to polluting coal and nuclear stations.

Hence, it would be beneficial to inform and involve the public already in the planning phase of a wind energy project. Fears and preoccupations regarding the erection of the turbines should be taken seriously. Presenting achieved progress in the field of noise, giving background information about the whole project, establishing visitor centers, and "open days" could help to achieve a change in general attitude.

Opinion about the turbines may change as is reported in the study carried out by Exeter University about public acceptance near the Delabole wind farm in UK which increased from 17 % before construction of the wind farm to 85 % after the wind farm had commenced operation [200].

1.3.2 Visual Impact

The effect of a wind turbine on the landscape as experienced by an observer is called visual impact, see Figure 1.2. The more valuable the landscape, the more sensitive the public will be with regard to visual intrusion. According to Corbet [36], wind turbines are much more accepted in industrial or large-scale agriculture areas. In scenic areas or in smaller-scale landscapes, the machines can look intrusive and thus would be less welcome. According to Gipe [74]:

> The most significant means for improving public acceptance is by providing visual uniformity. Even when large numbers of turbines are concentrated in a single array, or there are several large arrays in one locale, visual uniformity can create harmony in an otherwise disturbing vista. Visual uniformity is simply another way of saying that the rotor, nacelle and tower of each machine look similar. They need not be identical.

Corbet [36] also mentions that a wind farm arranged in a line is preferred to that in a group or cluster, and tubular towers are considered to look more attractive than lattice towers. Gipe [73] suggests that

> the designers should strive toward visual unity between rotor, nacelle and tower. A wind turbine needs not be a "box on a stick". Lattice towers need not appear cluttered with cross braces and angular lines. Lattice towers can be designed with graceful curves and a sparing use of cross braces.

Another aspect is the amount of land used for a wind turbine. Gipe [73] recommends to minimize the number of roads in a wind park. This would prevent unsightly cut and fill slopes in steep terrain.

Figure 1.2: Visual impact [213].

Furthermore, Gipe [74] mentions that it is important to keep the turbines spinning at low wind speeds so that they are perceived as being useful and therefore beneficial. If the turbines have any operational problem there must be a fast reaction to repair them or to remove them, if there is any unrepairable fault. Otherwise, expectations of the observers will be disappointed and the public could regard them as disturbing the landscape.

However, Stevenson [197] reports that the opinion polls carried out in Europe and the USA suggest that people's perception of wind power as a clean and renewable source of energy outweighs the perceived visual impact.

1.3.3 Electromagnetic Interference

Any large structure containing substantial amounts of metal may cause interference with electromagnetic signals. The same is true in the case of wind turbines. Under normal atmospheric conditions, a signal emitted from a radio, communication links, television, or microwave senders reaches the receiver more or less undisturbed. In some instances, unwanted interference effects can occur with turbines that are positioned close to a receiver as exemplified in Figure 1.3.

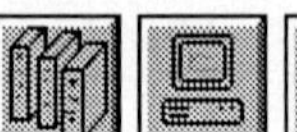

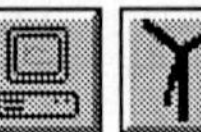

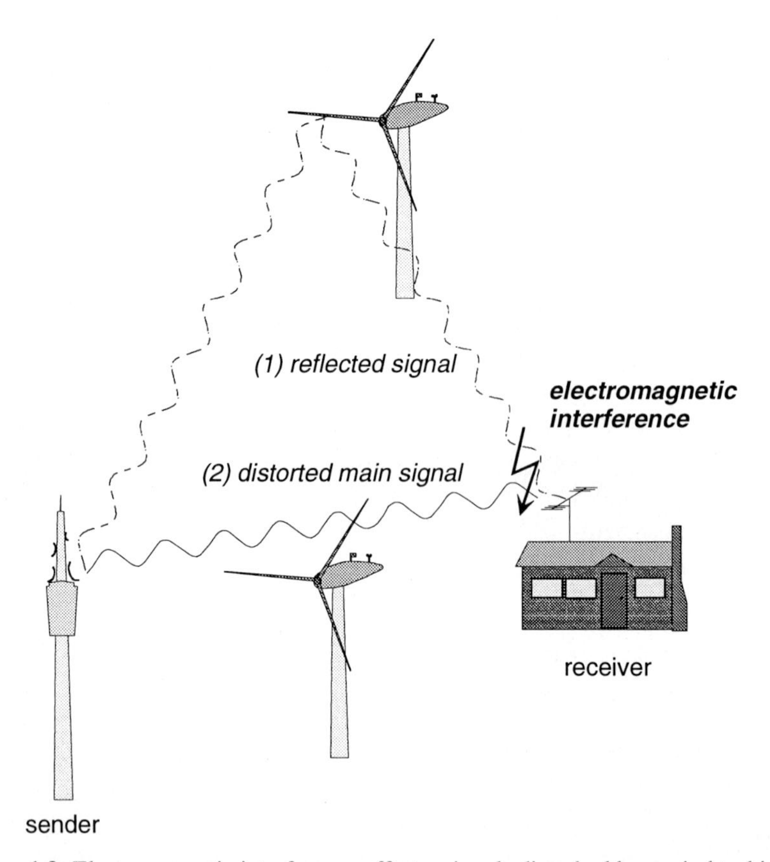

Figure 1.3: Electromagnetic interference effects: signals disturbed by a wind turbine.

Hau [92] gives a formula to estimate the distance from a turbine that is influenced by electromagnetic interference. Two possible effects have to be distinguished:

– The turbine is located on the connecting line between sender and receiver. The rotating blades cut through the waves and disturb the signal. This effect is most prominent in the UHF-band [92] which is used to transmit radio signals etc.
– The second effect is less significant and originates from the reflection of the original signal towards the receiver at the metal parts of the turbine. For TV signals this may cause secondary images.

The degree of electromagnetic interference caused by wind turbines varies depending on the following factors:

- *blade materials*: wooden blades absorb rather than reflect radio waves, whereas electromagnetic interference may occur if the turbine has metal blades or glass-reinforced plastic (GRP) blades containing metal components, for example, lightning protection,
- *surface shape of the tower*: smooth rounded towers reflect less than faceted towers.

Apart from electromagnetic waves from TV, those from microwave links, VHF Omni-directional Ranging, and Instrument Landing Systems (ILS) are most sensitive to interference. Thus, the installation of large machines around airports and other sensitive areas may be limited. Nevertheless, the problem of electromagnetic interference arises only under special conditions and especially for TV signals. It can be avoided by applying support senders, cable connections or by re-adjusting the antennas.

1.3.4 Additional Environmental Factors

Further parameters such as safety, shadow flicker, bird life, flora and fauna are also important factors to be considered while assessing the impact of wind turbines.

There are two ways in which birds could be affected by wind turbines. The first is direct collision of birds with turbines, and the second is the disturbance caused by turbines, which may lead to the reduction of their habitat. Stevenson [197] reports:

> Although studies in the Netherlands, Denmark, Sweden, the USA and the UK suggest that the chance of collision in daylight, with good visibility, is negligible, there is reason for caution where wind farms are situated on migration pathways or in habitats of rare birds.

A study conducted by F. Lubbers et al. [160] on the flying behavior of birds showed that

> during the daytime and also by night, birds generally detect the wind turbines in time to adapt their flying routes and avoid going too close to the moving rotor blades.

Although bird fatalities have been noticed, an extrapolation of the obtained data shows that bird fatalities per kilometer of a wind farm never exceed those per kilometer motorway or high-voltage overhead line.

  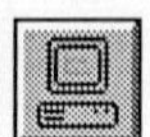 

1.4 Road Map of the Book

The previous sections of this chapter cover some of the background information and a short discussion of aspects relevant for the understanding of the wind turbine noise problem. Figure 1.4 shows wind turbine noise assessment factors and some of the main aspects covered within the book, i.e. the generation of noise, its propagation through the atmosphere and effects resulting at the receiver location. The structure of the book will be described in the following paragraphs.

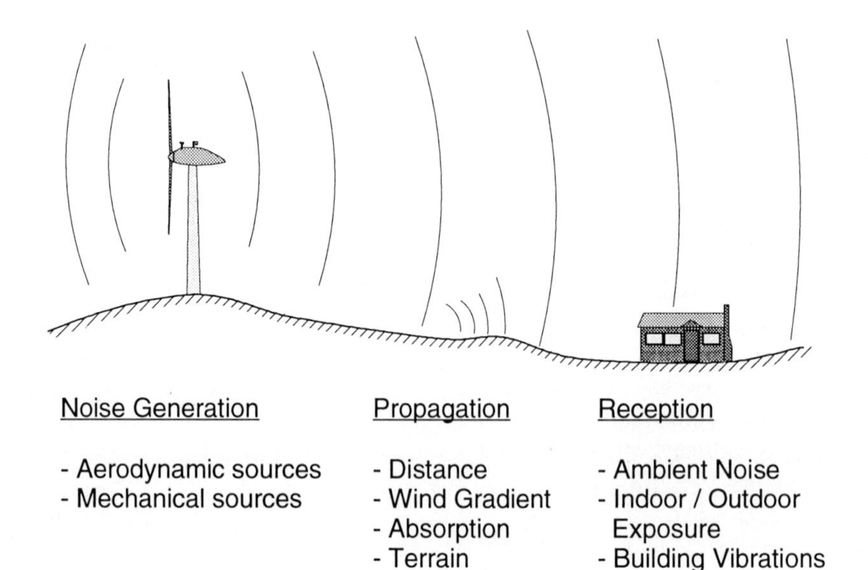

Figure 1.4: Wind turbine noise assessment factors [190].

Chapter 2 introduces the basic definitions which are used to describe sound and noise, as sound pressure level, sound power level, directivity, and different types of spectra.

Chapter 3 explains the theoretical background of (aero-)acoustics to the more academically interested reader. However, the following chapters can be fully understood without reading Chapter 3. It is intended to serve as a starting point for people interested more in acoustic theories. The chapter contains sections dealing with the fundamental wave equation, elementary sound

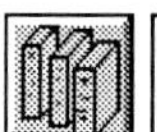
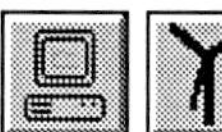

sources, and the generation of aerodynamic sound by turbulence and its interaction with solid surfaces.

Chapter 4 describes the noise generating mechanisms of wind turbines for low and high frequencies. Since mechanical noise is of less significance, mainly aerodynamic noise will be discussed. Typical noise spectra are shown together with a discussion of the separate mechanisms and a discussion of their relative importance.

Chapter 5 gives a summary of rules of thumb and state-of-the-art noise prediction codes, together with the implemented formulas. Typical results demonstrate capabilities and limitations of the codes. The chapter concludes with a discussion of the various codes.

Chapter 6 explains the process of noise propagation through the atmosphere dealing with the underlying effects, such as air absorption, weather effects, ground effect, screening, and noise propagation in complex terrain. One of the most important prediction methods together with models recently developed in EU-financed projects are presented together with sample results.

Chapter 7 explains how noise and flow can be measured. Experimental facilities, as acoustic wind tunnels, are described together with a discussion of capabilities and limitations of experimental research in the field of noise. Different acoustic measurement techniques such as the ground board, the acoustic parabola, the proximity microphone, and the acoustic antenna are briefly presented, and finally a discussion of flow visualization techniques.

Chapter 8 discusses means for noise reduction of the noise mechanisms discussed in Chapter 6, including trailing-edge serrations, modification of trailing-edge shape and material, modification of blade tip shape, and noise reduction by change in rotation speed and pitch.

Chapter 9 gives a list of recommendations for research in the field of noise for the future. It is followed by a list of recommended references in Chapter 10 which could serve as a starting point for research in the field of wind turbine noise.

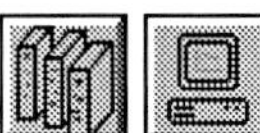
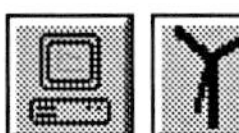

2 Noise and its Effects

2.1 Sound and Noise

Sound can be generated by a multitude of mechanisms. The sound emitted by a vibrating surface, e.g. a bell or a loudspeaker, is well-known to most people. Another mechanism is the periodic injection of air in the case of a siren. Several mechanisms are involved in the production of aerodynamic noise from wind turbines which is the topic of this book. Chapters 3–5 are dedicated to this subject.

Sound is always associated with rapid small-scale *pressure* fluctuations overlying the normal atmospheric pressure. These fluctuations are emitted from a *source* and travel as *waves* through the medium at the *speed of sound* (approximately 340 m/s in air). Sound may be reflected, partially absorbed, or attenuated before reaching the human eardrum where it produces a sensation of hearing depending on the amplitude of the sound wave.

In general, one could say that sound turns into *noise* in case it is unwanted. Whether sound is perceived as noise or not depends on subjective factors such as the sensitivity of the listener and the situation, but also on measurable quantities like the level and duration. A dripping faucet produces a relatively low sound level but can make it impossible for people to sleep. On the other hand, many people can sleep in a car, train, or even airplane where enormous sound levels can occur.

Several physical quantities have been defined which enable to compare and classify different sounds and which also give indications for the human perception of noise. This chapter deals with the definition and interpretation of these quantities and gives an overview of the regulations which most countries introduced in order to minimize the impact of noise on people.

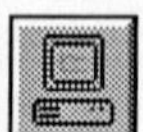

2.2 Definitions

2.2.1 Sound Pressure *p*

Sound is characterized by small pressure fluctuations which overly to the atmospheric pressure. These fluctuations propagate through the medium as sound waves. The pressure perceived at an observer location is a function of time and is called the sound pressure $p(t)$.

The time signal of sound pressure may have an arbitrary shape. However, it can always be characterized as a superposition of harmonic signals of different frequency and amplitude. Such harmonic signals are the basic patterns of sound and are called pure tones.

2.2.2 Amplitude *A*, Frequency *f*, Period *T*, Angular Frequency ω

The sound pressure signal $p(t)$ of a pure tone is specified by its amplitude A and its frequency f

$$p(t) = A \cdot \cos(2\pi f \cdot t) = A \cdot \cos\left(2\pi \cdot \frac{t}{T}\right) = A \cdot \cos(\omega \cdot t) . \qquad (2.1)$$

f denotes the number of cycles per second and is measured in Hertz, Hz. The inverse of the frequency f is the period T, which is the duration of one cycle. Multiples of Hz are used for high frequencies: 1 kHz = 10^3 Hz. The amplitude A is measured in Pascal, Pa. The angular frequency ω is defined as 2π times the frequency f

$$f = \frac{1}{T} \qquad \omega = 2\pi f = \frac{2\pi}{T} . \qquad (2.2)$$

A normal ear is sensitive to tones with frequencies ranging from about 16 Hz to 16000 Hz. Ordinary speech is concentrated in a relatively narrow frequency band in the vicinity of 1000 Hz. Tones with frequencies below 16 Hz are called infrasound and above 16000 Hz ultrasound.

2.2.3 Sound Waves, Wave Speed c_0, Wavelength λ, Wave Number k_0

The small pressure fluctuations that constitute sound propagate as waves through a medium. These waves can be longitudinal or lateral depending on whether the motion of the particles is parallel or perpendicular to the direction of propagation. Lateral waves require that the medium can sustain lateral

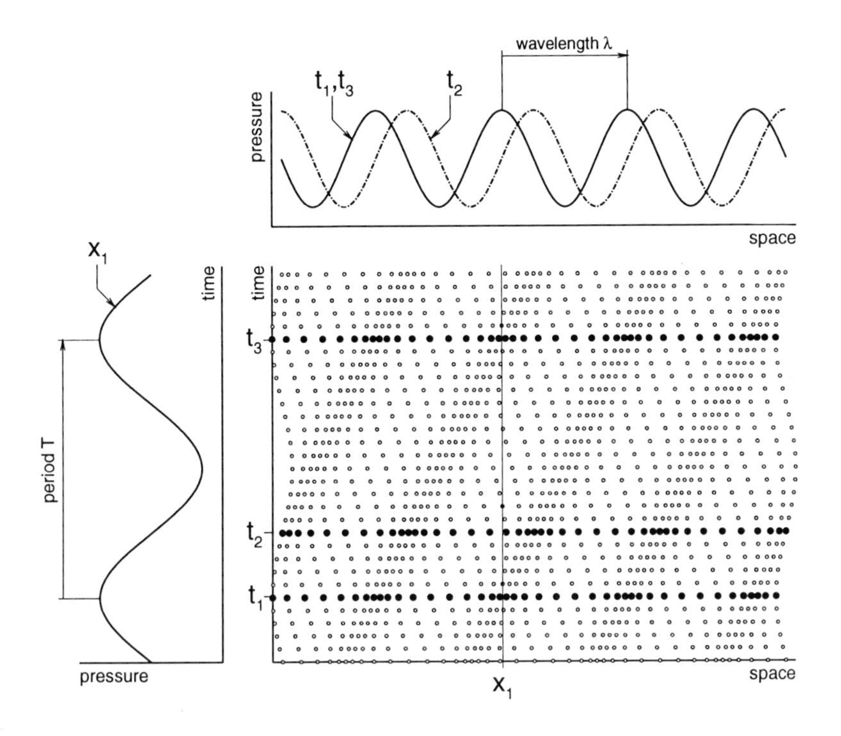

Figure 2.1: Illustration of sound wave, wavelength and period.

(shear) strains and therefore occur only in solids. In fluids, sound propagates in longitudinal waves. In the direction of propagation the medium becomes periodically denser and rarer. The latter corresponds to higher and lower sound pressure, respectively. The periodic change in density and pressure requires also a periodic ('back and forth') motion of fluid particles.

Figure 2.1 illustrates how a wave is propagating. The patterns of small particles show regions of high and low density which correspond to high and low sound pressure, respectively. The graph on top shows the spatial distribution of sound pressure for different points in time. It can be seen that these patterns are moving with time to the right ($t_1 \rightarrow t_2$). After a time T, which corresponds to the period of the sound pressure signal as shown in the left graph, the wave has moved by one wavelength to the right. The wavelength λ is the distance between two crests or troughs in the distribution of sound pressure. Thus, the velocity of the wave is

$$c = \frac{\lambda}{T} = f\,\lambda = \frac{\omega}{k} \tag{2.3}$$

which is the basic relationship between propagation speed, frequency, and wavelength. The wave number k is defined analogously to the angular frequency ω as the inverse of the wavelength λ times 2π

$$k = \frac{2\pi}{\lambda}. \tag{2.4}$$

Mathematically, the motion of the wave in Figure 2.1 can be expressed as a harmonic function of time and space

$$p(x,t) = A \cdot \cos\left(2\pi \cdot \left(\frac{x}{\lambda} \pm \frac{t}{T}\right)\right) = A \cdot \cos(kx \pm \omega t). \tag{2.5}$$

The – and + signs in equation (2.5) represent a wave moving in the positive and negative x-direction, respectively. Equation (2.5) is a solution of the linear wave equation (see section 3.3). Waves are common phenomena in all fields of physics. In case acoustic waves are considered, the acoustic wave speed is labeled c_0 and the acoustic wave number k_0. c_0 is often referred to as speed of sound and is typical for the medium in which the wave propagates. In dry air of 20°C c_0 has a value of 343 m/s. Table 2.1 gives the speed of sound in different media.

2.2.4 Sound Pressure Level L_p

The human ear does not response linearly to the amplitude of sound pressure. Doubling the amplitude produces the sensation of a somewhat louder sound, but it seems to be far less than twice as loud. For this reason, the scale customarily used to characterize sound pressure amplitudes is logarithmic, which is an approximation of the actual response of the human ear. The definition of the sound pressure level L_p is

$$L_p = 10 \cdot \log_{10}\left(\frac{\hat{p}^2}{\hat{p}_{\text{ref}}^2}\right) \tag{2.6}$$

where $\hat{p}$ is the root mean square sound pressure defined by

$$\hat{p}^2 = \lim_{T\to\infty}\left(\frac{1}{T}\int_0^T p^2(t)\mathrm{d}t\right) \tag{2.7}$$

and $\hat{p}_{\text{ref}}$ has the value of $2\cdot10^{-5}$ Pa. $\hat{p}_{\text{ref}}$ is the standard reference pressure corresponding to the weakest audible sound – the threshold of human hearing – at a frequency of 1000 Hz. L_p is expressed in decibel, dB.

Table 2.1: Speed of sound in different media.

	Medium	*Speed of sound c_0 (m/s)*
Gases (0°C)	Carbon dioxide	259
	Air, 20°C	343
	Helium	965
Liquids (25°C)	Ethyl alcohol	1207
	Water, pure	1498
	Water, sea	1531
Solids	Lead	1200
	Wood	~4300
	Iron and steel	~5000
	Aluminium	5100
	Glass, Pyrex	5170
	Granite	6000

Since the integration time in equation (2.7) is always finite, the equivalent continuous sound pressure level $L_{\mathrm{eq},T}$ is defined as

$$L_{\mathrm{eq},T} = 10 \cdot \log\left(\frac{1}{T}\int_0^T \frac{p^2(t)}{\hat{p}_{\mathrm{ref}}^2}\,\mathrm{d}t\right). \tag{2.8}$$

0 dB is the threshold of hearing. Doubling the amplitude produces an increase of 6 dB. Ten times the amplitude is a difference of 20 dB. Every increase of about 10 dB will result in doubling the sound's apparent loudness. Table 2.2 shows a range of the decibel-scale with typical examples.

2.2.5 Intensity I, Impedance Z_0, Sound Power P, Sound Power Level L_p

Like electromagnetic waves, sound waves transport energy from a source of sound through the medium. The sound intensity I is defined as the energy transmitted per unit time and unit area or the power per unit area. Far from a source of sound, the intensity is simply the mean square sound pressure (see equation 2.7) divided by the speed of sound and the density of the medium

$$I = \frac{\hat{p}^2}{\rho_0 c_0}. \tag{2.9}$$

$\rho_0 c_0$ is labeled the specific acoustic impedance Z_0

$$Z_0 = \rho_0 c_0 = 416\ \frac{\mathrm{kg}}{\mathrm{m}^2\mathrm{s}}\ \text{in air}. \tag{2.10}$$

Table 2.2: Decibel scale and examples [18].

Source	*Distance from the source (m)*	*Sound pressure level at ear position (dB(A))*
Threshold of pain		140
Ship siren	30	130
Jet engine	≈ 100	120
Thunder, artillery	≈ 1000	110
Steel riveter	4.5	100
Inside noisy factory		90
Inside tube train (open window), sport car		80
Inside average factory		70
Vacuum cleaner	3	
Freeway	30	
Loud conversation		60
Inside average office		50
Inside average living room		40
Soft whisper		30
Sound studio, quiet bedroom		20
Soundproof room		10
Threshold of hearing		0

ρ_0 is the undisturbed air density. Using the value of Z_0 for air, the sound pressure level L_p can also be expressed as the ratio of the intensity I and a reference intensity I_{ref}

$$L_p = 10 \cdot \log_{10}\left(\frac{\hat{p}^2}{\hat{p}_{ref}^2}\right) = 10 \cdot \log_{10}\left(\frac{I}{I_{ref}}\right) \tag{2.11}$$

where $I_{ref} = \hat{p}_{ref}^2 / \rho_0 c_0$ has the value of 10^{-12} W/m^2. Doubling of intensity yields an increase of 3 dB. Ten times the intensity is a difference of 10 dB.

Sound pressure level and intensity are properties of a field position. The total strength of a source of sound is characterized by the sound power emitted by the source. In general, the sound power transmitted through a surface S is the integral of the intensity I over S

$$P = \int_S I \, \mathrm{d}S . \tag{2.12}$$

If the surface S *encloses* the source of sound, P is the total sound power emitted by the source. The definition of the sound power level L_W is

$$L_W = 10 \cdot \log_{10}\left(\frac{P}{P_{ref}}\right) \tag{2.13}$$

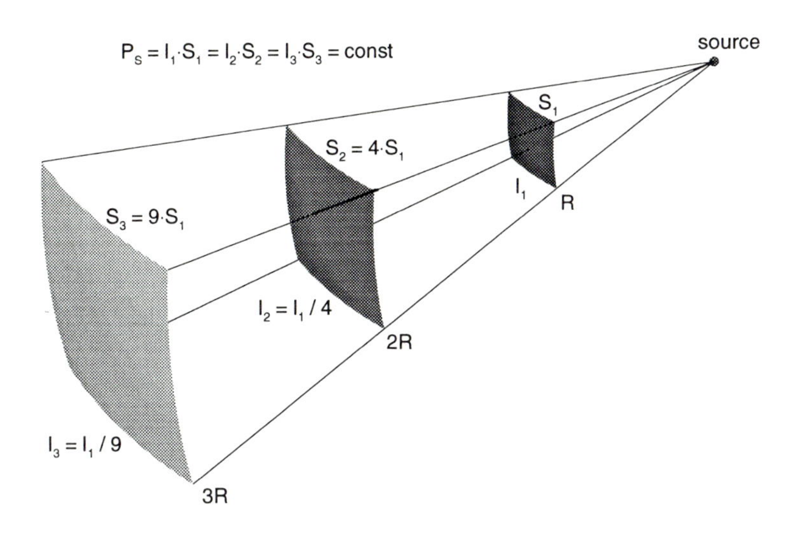

Figure 2.2: Illustration of the law of spherical spreading.

where $P_{ref} = 10^{-12}$ W is the standard reference sound power. The sound power level is again measured in decibel, dB. The eardrum can detect sound if the incoming sound power is as weak as 1 picowatt (10^{-12} W). If the ear is exposed to a power of more than 1 W, temporary hearing loss will result and contribute to permanent hearing loss.

In a free field without viscous damping, the acoustic intensity and the sound pressure decrease with the inverse square of the distance r. This *law of spherical spreading* is illustrated in Figure 2.2. The acoustic power P_S, transmitted through a surface S_1, is the same for the surfaces S_2 and S_3. If the intensity I is assumed to be constant over these surfaces, P_S is equal to I times the area of the surfaces (see equation (2.12)). Since the latter increases with the square of the radius r it follows

$$I = \frac{P_S}{4\pi r^2}. \tag{2.14}$$

Thus, a doubling in distance reduces the intensity to a quarter and the sound pressure level L_p by 6 dB.

It is important to distinguish between sound pressure level L_p and sound power level L_W. L_p is a property of sound *at a given observer location* and can be measured by a single microphone at that location. L_W is a property of the *source of sound* as it gives the total acoustic power emitted by the source. Since the sound pressure level normally depends on the direction of the

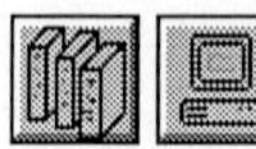

observer relative to the source, the determination of L_W requires combined measurements of L_p at several locations around the source. Sound pressure level and sound power level are often commonly labeled as sound level.

2.2.6 Narrow-band, 1/3- and 1/1-Octave Spectra

To characterize a source of sound, the frequency spectrum of the emitted sound has to be determined. The spectrum indicates which frequencies are prevalent in a sound pressure signal. It can reveal whether there are tonal components or a broadband swishing. The determination of a spectrum is called a frequency analysis. The frequency range of the spectrum is divided into several bands. The sound level is determined for each band by using filters which cancel out all frequencies outside the band. Three types of bands are commonly used: narrow bands, 1/3-octave bands, and 1/1-octave bands.

For a narrow-band spectrum, each frequency band has the same width Δf. Such a spectrum gives the most detailed picture of a sound signal. It is useful if the signal contains strong periodic or tonal components as, for example, propeller noise. Note, that the bandwidth Δf has to be mentioned explicitly. 1/3-octave and 1/1-octave spectra are used to characterize broadband signals that contain no prevalent frequencies. The terms 1/1-octave and 1/3-octave describe the frequency band according to the ratio of the upper and lower frequencies that bound the band.

In a 1/1-octave band, the upper bounding frequency is double the lower bounding frequency. In music, such an interval is called an octave because it contains eight notes. The center frequencies and the upper and lower bounds are defined in Table 2.3.

In a 1/3-octave band, the upper frequency is $\sqrt[3]{2}$ times the lower frequency. Each 1/1-octave band is made up of three equal, contiguous 1/3-octave bands. The corresponding formulas are given in Table 2.3. Since the frequencies computed with these formulas are odd numbers, the center frequencies are rounded according to ISO 266 [115]. Figure 2.3 shows examples for narrow-band, 1/3-octave, and 1/1-octave spectra.

A spectrum with a higher resolution can be transformed into a spectrum of lower resolution but not vice versa. To obtain the sound level for a frequency band all single sound intensities which lie in that band are added up

$$L_{\text{sum}} = 10 \cdot \log_{10}\left(\sum_{i=1}^{n} 10^{0.1 \cdot L_i} \right). \tag{2.15}$$

The ear is not equally sensitive to tones of different frequencies. Maximum response occurs for tones between 3000 and 4000 Hz, where the threshold for hearing is somewhat less than 0 dB. A 100 Hz tone must have an intensity level of at least 40 dB to be heard.

Therefore, the A-, B-, and C-weighted sound levels have been introduced where the lower frequencies are de-emphasized in a manner similar to human hearing (see Figure 2.4). A-weighting is most commonly used and is well suited for not too high levels. However, if strong low-frequent sound levels occur, B- or C- weighting is more appropriate. Weighted sound levels are also measured in decibels. The letters A, B, or C in brackets indicate whether A-, B-, or C-weighting has been used. An A-weighted sound level is measured in dB(A), the sound pressure level is labeled L_{pA}, and the sound power level L_{WA}.

A frequency analysis can be performed by means of a sound meter that can be tuned to different parts of the frequency range. The meter eliminates or filters out all sound components except those in a selected band of frequencies. Thus it is possible to measure selectively the sound level for different bands and to describe the frequency distribution as a set of partial sound levels in contiguous frequency bands covering the entire audible range.

The second possibility of performing a frequency analysis is to record periodic samples of a signal on tape and to analyze the signal afterwards, employing DFT (discrete Fourier transform) or FFT (fast Fourier transform) routines. However, care must be taken to choose the proper sampling rate in order to distinguish all frequencies properly and to avoid contamination of the spectrum which may result from the finite measurement time. An introduction to the problems of digital signal analysis can be found in [19], [194].

2.2.7 Emission, Immission, Directivity

Emission refers to sound emitted by a source, whereas immission refers to sound perceived by an observer. Emission, being a property of the source, is characterized by the sound power level L_W. Immission, being also a property of the observer position, is characterized by the sound pressure level L_p.

Most sound sources do not radiate uniformly in each direction. Therefore, the perceived sound pressure level does not only depend on the sound power level of the source and the distance but also on the directivity pattern which is typical for each sound source. Examples of directivity patterns are shown in Figure 5.5 (Section 5.2).

2.2.8 Annoyance

Annoyance cannot be determined from a measurement of the physical quantities of perceived noise solely, because it depends on the listeners subjective perception.

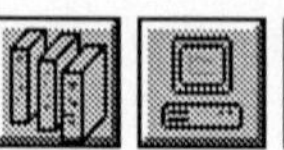

Table 2.3: Definition of 1/3- and 1/1-octave bands; A-, B-, and C-weighting.

Nominal frequency (Hz)				*Exact center frequency (Hz)*	*A-weighting (dB)*	*B-weighting (dB)*	*C-weighting (dB)*
1/3-octave		*1/1-octave*					
i		*j*					
-20	10			10,00	-70,4	-38,2	-14,3
-19	12,5			12,59	-63,4	-33,2	-11,2
-18	16	-6	16	15,85	-56,7	-28,5	-8,5
-17	20			19,95	-50,5	-24,2	-6,2
-16	25			25,12	-44,7	-20,4	-4,4
-15	31,5	-5	31,5	31,62	-39,4	-17,1	-3,0
-14	40			39,81	-34,6	-14,2	-2,0
-13	50			50,12	-30,2	-11,6	-1,3
-12	63	-4	63	63,10	-26,2	-9,3	-0,8
-11	80			79,43	-22,5	-7,4	-0,5
-10	100			100,0	-19,1	-5,6	-0,3
-9	125	-3	125	125,9	-16,1	-4,2	-0,2
-8	160			158,5	-13,4	-3,0	-0,1
-7	200			199,5	-10,9	-2,0	0,0
-6	250	-2	250	251,2	-8,6	-1,3	0,0
-5	315			316,2	-6,6	-0,8	0,0
-4	400			398,1	-4,8	-0,5	0,0
-3	500	-1	500	501,2	-3,2	-0,3	0,0
-2	630			631,0	-1,9	-0,1	0,0
-1	800			794,3	-0,8	0,0	0,0
0	1000	0	1000	1000,0	0,0	0,0	0,0
1	1250			1259	0,6	0,0	0,0
2	1600			1585	1,0	0,0	-0,1
3	2000	1	2000	1995	1,2	-0,1	-0,2
4	2500			2512	1,3	-0,2	-0,3
5	3150			3162	1,2	-0,4	-0,5
6	4000	2	4000	3981	1,0	-0,7	-0,8
7	5000			5012	0,5	-1,2	-1,3
8	6300			6310	-0,1	-1,9	-2,0
9	8000	3	8000	7943	-1,1	-2,9	-3,0
10	10000			10000	-2,5	-4,3	-4,4
11	12500			12590	-4,3	-6,1	-6,2
12	16000	4	16000	15850	-6,6	-8,4	-8,5
13	20000			19950	-9,3	-11,1	-11,2

Band	*Center frequency*	*Lower bound*	*Upper bound*
1/3-octave	$1000 \cdot 10^{0.1 \cdot i}$	$1000 \cdot 10^{0.1 \cdot (i-0.5)}$	$1000 \cdot 10^{0.1 \cdot (i+0.5)}$
1/1-octave	$1000 \cdot 10^{0.3 \cdot j}$	$1000 \cdot 10^{0.3 \cdot (j-0.5)}$	$1000 \cdot 10^{0.3 \cdot (j+0.5)}$

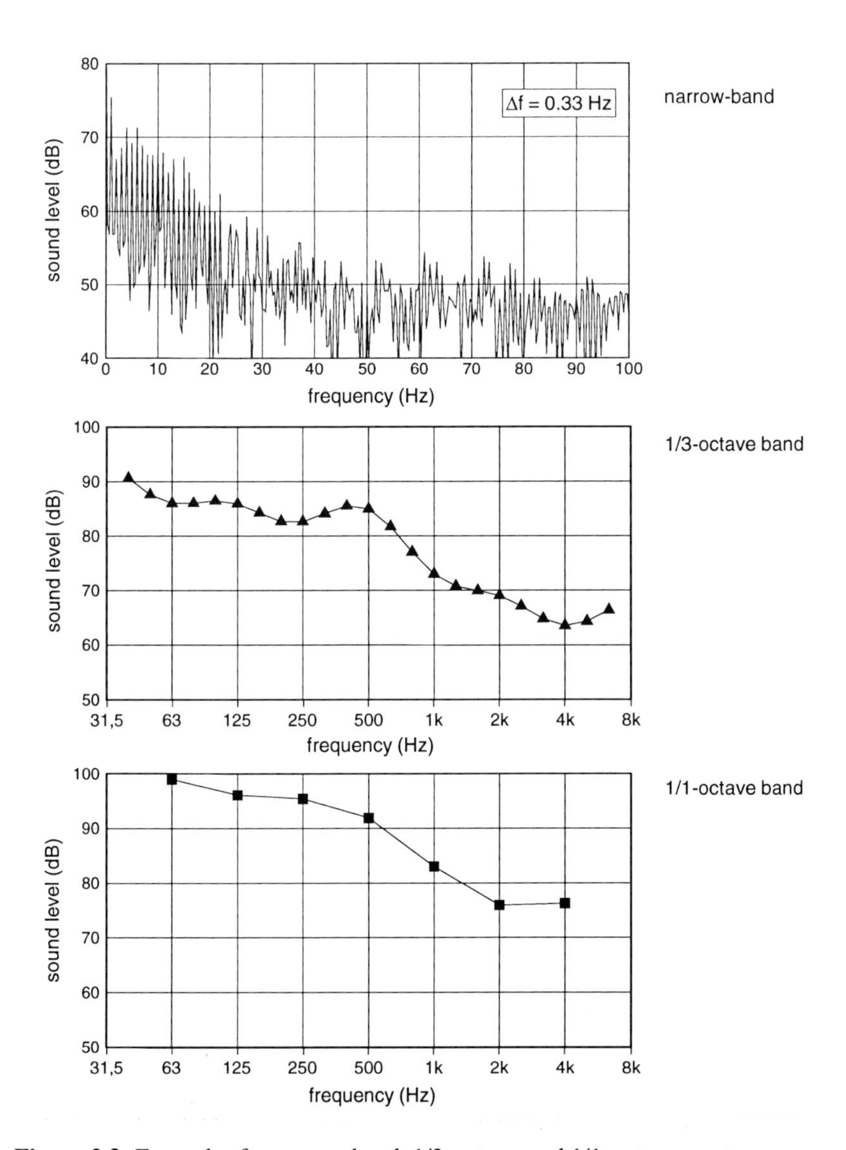

Figure 2.3: Examples for narrow-band, 1/3-octave, and 1/1-octave spectra.

Nevertheless, attempts have been made to quantify annoyance. Schultz completed a comprehensive study [187]. From the 19 conducted attitude surveys of people exposed to noise, he deduced a universal response curve relating the percentage of persons reporting to be highly annoyed (%H.A.) by noise to the noise level

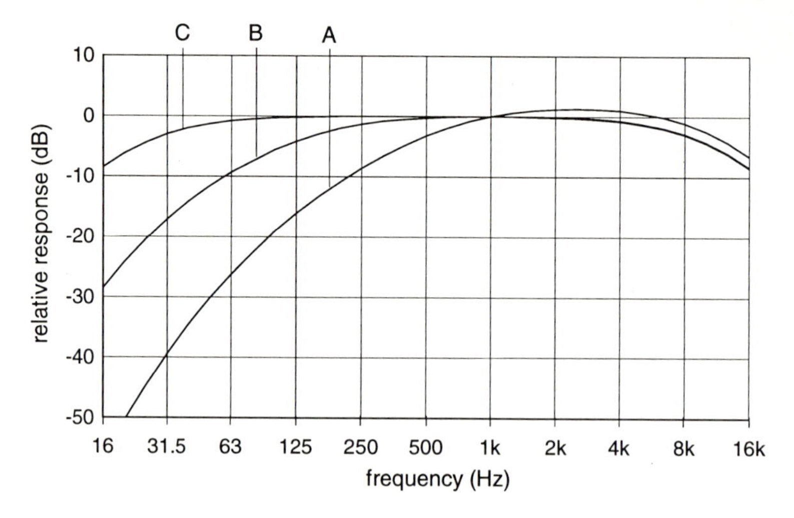

Figure 2.4: Definition of A-, B-, and C-weighting.

$$\%\text{H.A.} = 0.8553\, L_{dn} - 0.0401\, L_{dn}^2 + 0.00047\, L_{dn}^3 \tag{2.16}$$

where L_{dn} is an energy-average measure of the A-weighted noise intrusion, with a penalty applied for night-time noise intrusions (see also Figure 2.5).

2.3 Noise Regulations

In most countries, noise regulations define upper bounds for the noise to which people may be exposed. These limits depend on the areas and are different for daytime and night-time. For example, in Germany, a sound power level of 60 dB(A) is acceptable in commercial areas but not in residential areas. 55 dB(A) is acceptable in residential areas during the day but not during the night when 40 dB(A) is the limit.

All noise regulations provide a penalty of typically 5 dB(A) for pure tones. Thus, if a wind turbine produces a sound power level of 40 dB(A), which might be below the limit, but also a strong whistling, 5 dB(A) are added which forces the manufacturer to reduce the sound pressure level to 35 dB(A) or to remove the whistling tone. Table 2.4 gives an overview of the noise limits in three European countries.

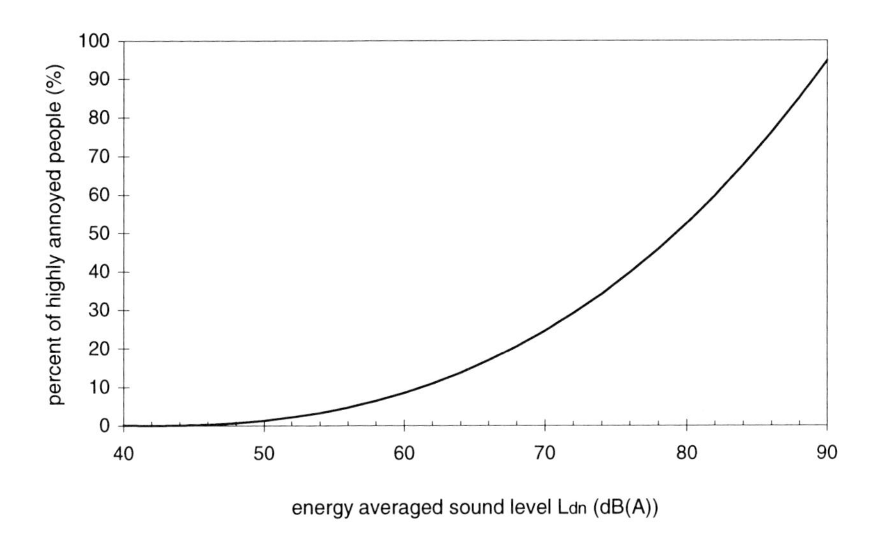

Figure 2.5: Noise level and number of annoyed people [187].

Table 2.4: Noise limits for equivalent sound pressure levels L_{Aeq} (dB(A)) in different European countries [75].

Country	*Commercial*	*Mixed*	*Residential*	*Rural*
Denmark			40	45
Germany				
– day	65	60	55	50
– night	50	45	40	35
Netherlands				
– day		50	45	40
– night		40	35	30

  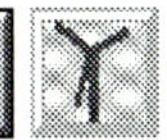

3 Introduction to Aeroacoustics

3.1 Introduction

One aim of this book is to give suggestions how aerodynamic noise from wind turbines can be reduced in order to increase the public's acceptance of this energy source. An important prerequisite to find solutions that go beyond state-of-the-art knowledge is to understand the mechanisms of aerodynamic sound generation. Therefore, this chapter gives an introduction to some basic equations and concepts of aeroacoustics.

The starting point is the linear wave equation, which is derived from the basic conservation laws of fluid dynamics. The elementary solutions of the wave equation – monopoles and dipoles – are described. Lighthill's acoustic analogy introduces the important concept of quadrupole radiation which is a by-product of all turbulent flows [22], [139], [140], [141], [143]. After a short description of Powell's theory of vortex sound [22], [179] the problem of solid surfaces within the field is tackled [22], [37], [65]. The chapter ends with a discussion of some important theories for sound generation due to lifting surfaces [3], [22], [66], [106]. The latter form the basis for the noise prediction schemes which are introduced in Chapter 5. An overview of different aspects of aeroacoustics can be found in [22], [23], [67], [141], [178].

3.2 Definitions

3.2.1 Harmonic Functions in Time and Space

In Chapter 2, the basic definitions which are needed to describe sound pressure signals and sound waves have been introduced; namely period T, frequency f, angular frequency ω, wavelength λ, wave number k, and wave speed c. Note that k_0 denotes the acoustic wave number and c_0 the acoustic

 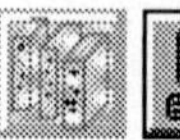 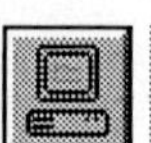 

wave speed (speed of sound). Furthermore, the basic relationship between frequency, wavelength and wave speed has been derived (equation (2.3)).

The harmonic functions which are used to describe the sound pressure signal of a pure tone (equation (2.1)) or the motion of a harmonic wave (equation (2.5)) can be rewritten by using the complex notation

$$p'(t) = A \cdot e^{-i\left(2\pi \frac{t}{T}\right)} = A \cdot e^{-i \cdot \omega t} \tag{3.1}$$

and

$$p'(x,t) = A \cdot e^{i\left(2\pi \frac{x}{\lambda} - 2\pi \frac{t}{T}\right)} = A \cdot e^{i(kx - \omega t)} \tag{3.2}$$

where $i = \sqrt{-1}$. This notation will be applied throughout this chapter. It is common to use $-i$ for the time dependence in equation (3.1). Equation (3.2) describes a one-dimensional *wave* moving in the positive x-direction. In two- or three-dimensional space, a wave $p'(\vec{x},t)$ can move in an arbitrary direction $\vec{n}$ where $|\vec{n}| = 1$. $p'(\vec{x},t)$ can be expressed by introducing the wave vector $\vec{k}$ which is aligned with $\vec{n}$

$$p'(\vec{x},t) = A \cdot e^{i\left(\vec{k}\vec{x} - \omega t\right)} = A \cdot e^{i\left(k_1 x_1 + k_2 x_2 + k_3 x_3 - \omega t\right)}. \tag{3.3}$$

The wave vector $\vec{k}$ is defined analogously to the one-dimensional wave number k in equation (2.4)

$$\vec{k} = \frac{2\pi}{\lambda} \vec{n} = \begin{pmatrix} k_1 & k_2 & k_3 \end{pmatrix} = \begin{pmatrix} \frac{2\pi}{\lambda_1} & \frac{2\pi}{\lambda_2} & \frac{2\pi}{\lambda_3} \end{pmatrix} \tag{3.4}$$

where λ_i is the wavelength in the x_i-direction. Figure 3.1 illustrates the definition of $\vec{k}$ in the x_1x_2-plane (k_3 is zero).

3.2.2 Fourier Transforms in Time and Space

The importance of harmonic functions is based on the fact that each non-harmonic function g of an independent variable t, x and $\vec{x}$ (or a combination) can be expressed as the sum of an infinite series of harmonic functions with different angular frequencies ω, wave numbers k, and wave vectors $\vec{k}$, respectively. The amplitudes of these harmonic functions are called the spectrum of the function g. The mathematical tool for the computation of the spectrum is the Fourier transform. An introduction can be found in [144].

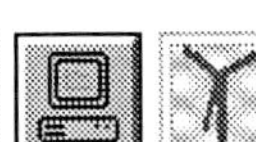

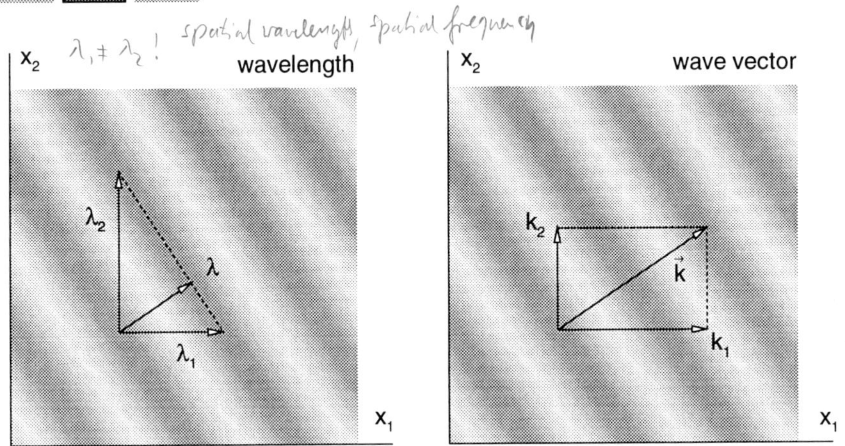

Figure 3.1: Illustration of wave vector $\vec{k}$.

If the function $g(t)$ depends on time t, the Fourier transform $\tilde{g}(\omega)$ is given by the Fourier integral

$$\tilde{g}(\omega) = \frac{1}{2\pi} \int_{-\infty}^{+\infty} g(t) \cdot e^{+i\omega t} \mathrm{d}t \,. \tag{3.5}$$

In a figurative sense, the operation defined by equation (3.5) "checks" the agreement between the function $g(t)$ and harmonic functions $e^{i\omega t}$ with different ω. If the agreement is good, the respective value of $\tilde{g}(\omega)$ is large, if not, $\tilde{g}(\omega)$ is small. The inverse Fourier transform yields the original signal $g(t)$

$$g(t) = \int_{-\infty}^{+\infty} \tilde{g}(\omega) \cdot e^{-i\omega t} \mathrm{d}\omega \,. \tag{3.6}$$

Note that the dimension of $\tilde{g}(\omega)$ is equal the dimension of $g(t)$ times s^{-1}. As shown in [144], a Fourier series is the special case of a Fourier transform for periodic signals. A pair of Fourier transforms can be defined for space dependence $g(x)$ as well. The kernel $e^{i\omega t}$ in equation (3.5) with the angular frequency ω is then replaced by e^{ikx} with the wave number k. If the function $g(\vec{x},t)$ depends on space *and* time, the kernel is replaced by the expression $e^{i(\vec{k}\vec{x}-\omega t)}$ for a plane wave traveling in the $\vec{n}$ direction with angular frequency ω and wave vector $\vec{k}$. Now, the Fourier transform "checks" the agreement of the function $g(\vec{x},t)$ with plane waves of different $\vec{k}$ and ω. This Fourier

transform is called a space–time transform and the corresponding spectrum is labeled a wave vector frequency ($\vec{k}\omega$) spectrum. The $\vec{k}\omega$ spectrum of the pressure fluctuations on a surface is used in Howe's model for trailing edge noise [106] (see Section 3.7). Special problems arise if the function g is not continuous but given by a set of discrete samples which have been obtained from measurements or simulations. A good introduction to the field of discrete and fast Fourier transforms can be found in [19], [194].

3.2.3 Generalized Functions

Generalized functions are well-suited for many applications in acoustic and electromagnetic theory. An introduction can be found in [144]. The most important generalized function is the delta function, which can be defined as

$$\delta(x) = \lim_{\varepsilon \to 0} \begin{Bmatrix} 0 & \text{if} \ |x| > \varepsilon/2 \\ \frac{1}{\varepsilon} & \text{if} \ |x| < \varepsilon/2 \end{Bmatrix}. \tag{3.7}$$

$\delta(x)$ is also called Dirac's function. $\delta(x)$ is zero for all x, except for x=0 where it approaches infinity. The integral over $\delta(x)$ is defined to be unity

$$\int_{-\infty}^{+\infty} \delta(x) \mathrm{d}x = 1. \tag{3.8}$$

Note from definition (3.7) and from equation (3.8) that the dimension of the delta function is the inverse of the dimension of the argument, e.g. $[\delta(x)] = \mathrm{m}^{-1}$ if $[x]$ = m. A delta function can be defined analogously in three dimensions

$$\begin{aligned} \delta(\vec{x}) &= \delta(x_1)\delta(x_2)\delta(x_3) \\ &= \lim_{\substack{\varepsilon_1 \to 0 \\ \varepsilon_2 \to 0 \\ \varepsilon_3 \to 0}} \begin{Bmatrix} 0 & \text{if} \ |x_1| > \frac{\varepsilon_1}{2}, \ |x_2| > \frac{\varepsilon_2}{2}, \ |x_3| > \frac{\varepsilon_3}{2} \\ \frac{1}{\varepsilon_1 \varepsilon_2 \varepsilon_3} & \text{if} \ |x_1| < \frac{\varepsilon_1}{2}, \ |x_2| < \frac{\varepsilon_2}{2}, \ |x_3| < \frac{\varepsilon_3}{2} \end{Bmatrix}. \end{aligned} \tag{3.9}$$

As above, $\delta(\vec{x})$ is zero for all $\vec{x}$ except $\vec{x} = \vec{0}$ (or $x_1 = x_2 = x_3 = 0$) where it approaches infinity. The integral over $\delta(\vec{x})$ is defined to be unity

$$\int\int\int_{-\infty}^{+\infty} \delta(\vec{x}) \mathrm{d}\vec{x} = \int_{-\infty}^{+\infty} \delta(x_1) \mathrm{d}x_1 \int_{-\infty}^{+\infty} \delta(x_2) \mathrm{d}x_2 \int_{-\infty}^{+\infty} \delta(x_3) \mathrm{d}x_3 = 1. \tag{3.10}$$

Again, if the arguments x_1, x_2, x_3 have the dimension m, the dimension of the three-dimensional delta function is m^{-3}. The three-dimensional delta function will be useful for the discussion of the elementary solutions of the wave equation in the following sections.

Another important generalized function is the Heaviside function $H(x)$ which is defined as

$$H(x) = \begin{cases} 0 & \text{if} \quad x < 0 \\ 1 & \text{if} \quad x > 0 \end{cases}. \tag{3.11}$$

$H(x)$ is also called the step function. $H(x)$ is the integral of $\delta(x)$, whereas $\delta(x)$ is the derivative of $H(x)$

$$H(x) = \int_{-\infty}^{x} \delta(a)\mathrm{d}a \tag{3.12}$$

$$\delta(x) = \frac{\mathrm{d}}{\mathrm{d}x} H(x). \tag{3.13}$$

3.2.4 Summation Convention

Most equations in this chapter are written by using the Cartesian-tensor summation convention, in which repeated indices are to be summed over all three coordinates. For example, the divergence of a vector field F_i can be expressed as

$$\nabla \cdot \vec{F} = \frac{\partial F_i}{\partial x_i} = \frac{\partial F_1}{\partial x_1} + \frac{\partial F_2}{\partial x_2} + \frac{\partial F_3}{\partial x_3}. \tag{3.14}$$

The Laplacian of a scalar field p' is written

$$\nabla^2 p' = \frac{\partial^2 p'}{\partial^2 x_i} = \frac{\partial^2 p'}{\partial x_i \partial x_i} = \frac{\partial^2 p'}{\partial^2 x_1} + \frac{\partial^2 p'}{\partial^2 x_2} + \frac{\partial^2 p'}{\partial^2 x_3}. \tag{3.15}$$

A third example is

$$\frac{\partial}{\partial x_j}(\rho u_i u_j) = \frac{\partial}{\partial x_1}(\rho u_i u_1) + \frac{\partial}{\partial x_2}(\rho u_i u_2) + \frac{\partial}{\partial x_3}(\rho u_i u_3). \tag{3.16}$$

Note that no summation over i has been made because this index occurs only once.

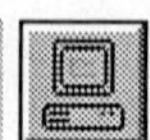

3.3 The Linear Wave Equation

The generation and propagation of aerodynamic sound is governed by the basic conservation laws of fluid dynamics: namely, the conservation of mass (continuity equation), the conservation of momentum (Euler- and Navier-Stokes equation for inviscid and viscous flows, respectively), and the conservation of energy (energy equation).

The starting point is the continuity equation and the momentum equation in the *i*th direction for an inviscid fluid (see e.g. [22])

$$\frac{\partial \rho}{\partial t}+\frac{\partial}{\partial x_i}(\rho u_i)=0 \tag{3.17}$$

$$\frac{\partial}{\partial t}(\rho u_i)+\frac{\partial}{\partial x_j}(\rho u_i u_j)+\frac{\partial p}{\partial x_i}=0\,. \tag{3.18}$$

The fluid is to be free of production of mass and external forces and body forces are neglected. The time-derivative of equation (3.17) and the divergence of equation (3.18) yield

$$\frac{\partial^2 \rho}{\partial t^2}+\frac{\partial^2}{\partial t \partial x_i}(\rho u_i)=0 \tag{3.19}$$

$$\frac{\partial^2}{\partial x_i \partial t}(\rho u_i)+\frac{\partial^2}{\partial x_i \partial x_j}(\rho u_i u_j)+\frac{\partial^2 p}{\partial x_i^2}=0\,. \tag{3.20}$$

The momentum density ρu_i can be removed by subtracting equation (3.20) from equation (3.19)

$$\frac{\partial^2 \rho}{\partial t^2}-\frac{\partial^2}{\partial x_i \partial x_j}(\rho u_i u_j)-\frac{\partial^2 p}{\partial x_i^2}=0\,. \tag{3.21}$$

The acoustic quantities ρ', p' and u_i' are defined as small departures from a state where the fluid is at rest with a uniform density ρ_0 and a uniform pressure p_0. ρ' is the acoustic density, p' the acoustic pressure, and u_i' the acoustic particle velocity in the *i*th direction

$$p=p_0+p', \quad \rho=\rho_0+\rho', \quad u_i=0+u_i'\,. \tag{3.22}$$

Introducing the definitions in equation (3.22) allows to linearise equation (3.21) by neglecting products of small quantities

$$\frac{\partial^2}{\partial t^2}(\rho_0+\rho')-\frac{\partial^2}{\partial x_i \partial x_j}\left((\rho_0+\rho')u_i'u_j'\right)-\frac{\partial^2}{\partial x_i^2}(p_0+p')=0 \tag{3.23}$$

$$\frac{\partial^2 \rho'}{\partial t^2}-\frac{\partial^2 p'}{\partial x_i^2}=0\,. \tag{3.24}$$

If the compression and expansion of the fluid is considered to be isentropic, there is a linear relationship between the acoustic pressure and the acoustic density

$$\begin{aligned} p-p_0 &= c_0^2(\rho-\rho_0) \\ p' &= c_0^2\rho' \end{aligned} \tag{3.25}$$

where c_0 depends on the fluid properties and is the velocity at which disturbances propagate through the medium (see Section 2.2.3). For perfect gases, c_0 is a function of the adiabatic gas constant κ, the specific gas constant R, and the thermodynamic temperature T

$$c_0^2=\left.\frac{\partial p}{\partial \rho}\right|_{S=\text{const}}=\kappa RT\,. \tag{3.26}$$

Introducing equation (3.25) into equation (3.24) yields the homogeneous wave equation for the acoustic density and the acoustic pressure:

$$\boxed{\frac{\partial^2 \rho'}{\partial t^2}-c_0^2\frac{\partial^2 \rho'}{\partial x_i^2}=0 \qquad \frac{1}{c_0^2}\frac{\partial^2 p'}{\partial t^2}-\frac{\partial^2 p'}{\partial x_i^2}=0}\,. \tag{3.27}$$

The two formulations of the wave equation are equivalent, since the acoustic pressure is always related to the acoustic density via equation (3.25). In the following, either the formulation with the acoustic density or the acoustic pressure will be used. The inhomogenous wave equation can be written as

$$\boxed{\frac{1}{c_0^2}\frac{\partial^2 p'(\vec{x},t)}{\partial t^2}-\frac{\partial^2 p'(\vec{x},t)}{\partial x_i^2}=\sigma(\vec{x},t)} \tag{3.28}$$

where the right-hand side $\sigma(\vec{x},t)$ represents arbitrary source terms which can be distributed throughout the field. The solutions of equation (3.28) for three elementary forcing terms are introduced in the next two sections.

The inhomogenous wave equation (3.28) is expressed in the *time domain*. For many applications it is helpful to use the Fourier transform in order to study the generation and propagation of sound in the *frequency domain*. This can be done by substituting each dependent variable by its Fourier integral

$$p'(\vec{x},t) = \int_{-\infty}^{+\infty} \tilde{p}'(\vec{x},\omega)\cdot e^{-i\omega t}\,\mathrm{d}\omega \qquad \sigma(\vec{x},t) = \int_{-\infty}^{+\infty} \tilde{\sigma}(\vec{x},\omega)\cdot e^{-i\omega t}\,\mathrm{d}\omega \tag{3.29}$$

Introducing equation (3.29) in equation (3.28) yields

$$\begin{gathered} \frac{1}{c_0^2}\frac{\partial^2}{\partial t^2}\left(\int_{-\infty}^{+\infty} \tilde{p}'(\vec{x},\omega)\cdot e^{-i\omega t}\,\mathrm{d}\omega\right) - \frac{\partial^2}{\partial x_i^{\,2}}\left(\int_{-\infty}^{+\infty} \tilde{p}'(\vec{x},\omega)\cdot e^{-i\omega t}\,\mathrm{d}\omega\right) = \\ \int_{-\infty}^{+\infty} \tilde{\sigma}(\vec{x},\omega)\cdot e^{-i\omega t}\,\mathrm{d}\omega \end{gathered} \quad . \tag{3.30}$$

Evaluation of the double time derivative leads to the wave equation in the frequency domain, which is termed a Helmholtz equation

$$\begin{gathered} \int_{-\infty}^{+\infty}\left(\frac{-\omega^2}{c_0^2}\tilde{p}'(\vec{x},\omega) - \frac{\partial^2}{\partial x_i^{\,2}}\tilde{p}'(\vec{x},\omega) - \tilde{\sigma}(\vec{x},\omega)\right)\cdot e^{-i\omega t}\,\mathrm{d}\omega = 0 \\ \frac{\partial^2 \tilde{p}'(\vec{x},\omega)}{\partial x_i^{\,2}} + k_0^2\tilde{p}'(\vec{x},\omega) = -\tilde{\sigma}(\vec{x},\omega) \end{gathered} \tag{3.31}$$

or

$$\nabla^2\tilde{p}' + k_0^2\tilde{p}' = -\tilde{\sigma}(\vec{x},\omega)\,. \tag{3.32}$$

The wave equation and the Helmholtz equation are important partial differential equations which govern many problems in physics. In principle, they can be solved with the method of Green's functions. The idea is to find a solution for an impulsive point source which is expressed as a delta function and forms the right-hand side of the differential equation. The solution for this special right-hand side is labeled the Green's function G. For example, the Green's function for the Helmholtz equation is defined by

$$\begin{gathered} \nabla^2 G + k_0^2 G = -\delta(\vec{x}-\vec{y}) \\ G = G(\vec{x},\vec{y},\omega) \end{gathered} \tag{3.33}$$

and is given by

$$G(\vec{x},\vec{y},\omega) = \frac{e^{\pm ik_0|\vec{x}-\vec{y}|}}{4\pi|\vec{x}-\vec{y}|}\,. \tag{3.34}$$

G is also called the free space Green's function. The selection of $+i$ or $-i$ determines if G describes an outward traveling or a collapsing wave (the

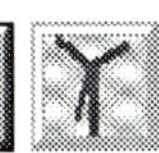

opposite is true if $+i$ is chosen for the time dependence in equation 3.1). Equation (3.34) is a solution of the Helmholtz equation only for the three-dimensional case. In two dimensions the Green's function of the Helmholtz equation is a Hankel function of zeroth order and first kind [79]. Once the Green's function is known, the solution of the inhomogenous differential equation can easily be found by integrating over the region where the right-hand side $\sigma(\vec{x},t)$ is non-zero

$$\tilde{p}'(\vec{x},\omega)=\int_V \tilde{\sigma}(\vec{y},\omega)\cdot G(\vec{x},\vec{y},\omega)\mathrm{d}V(\vec{y})\,. \tag{3.35}$$

Equation (3.35) is valid only in unbounded space. The influence of solid boundaries is tackled in Section 3.6. The methods for solving partial differential equations, especially the wave equation and the Helmholtz equation, have been studied in depth, e.g. by Morse and Feshbach [168].

In the far field of an arbitrary source of sound, the propagation of sound waves is approximately one-dimensional. Such waves are called *plane waves.* For a plane wave traveling in the x_1-direction, the wave equation (3.27) reduces to

$$\frac{1}{c_0^2}\frac{\partial^2 p'}{\partial t^2}-\frac{\partial^2 p'}{\partial {x_1}^2}=0\,. \tag{3.36}$$

The solution of equation (3.36) is given by

$$p'(x_1,t)=\tilde{A}_1 e^{i(k_0 x_1-\omega t)}+\tilde{A}_2 e^{i(k_0 x_1+\omega t)} \tag{3.37}$$

where k_0 and ω are related to c_0 via equation (2.3). The amplitudes A_1 and A_2 are complex quantities with an absolute value and a phase angle. The first term in equation (3.37) describes a wave traveling in the x_1-direction, the second term represents a wave traveling in the opposite direction (see also Section 2.2.3).

In the following, the so-called elementary solutions of the wave equation are introduced, namely *monopoles*, *dipoles*, and *quadrupoles.* Physically they can be identified with *sources of fluid*, *forces*, and *fluctuating Reynolds stresses*, respectively. The latter are present in all turbulent flows. This important result has been found by Lighthill [139], [141] and is referred to as Lighthill's acoustic analogy.

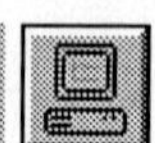

3.4 Elementary Solutions of the Wave Equation

In unconfined spaces all sound fields are three-dimensional and the description of the three-dimensional problem is therefore used to exemplify the role of sound sources.

3.4.1 Monopoles

If fluctuating sources of matter with an inflow-rate density $Q(\vec{x},t)$ are distributed in part of the medium, the continuity equation (3.17) has to be rewritten

$$\frac{\partial \rho}{\partial t}+\frac{\partial}{\partial x_i}(\rho u_i)=Q(\vec{x},t). \tag{3.38}$$

$Q(\vec{x},t)$ has the dimension kg/(m^{3}s). Performing the linearisation described before results in an additional term on the right-hand side of the wave equation

$$\frac{\partial^2 \rho'}{\partial t^2}-c_0^2\frac{\partial^2 \rho'}{\partial x_i{}^2}=\frac{\partial Q}{\partial t}=\dot{Q}(\vec{x},t), \tag{3.39}$$

which has the dimension kg/(m^{3}s^2). This term acts as a forcing term and gives rise to a non-trivial solution of the wave equation. For a concentrated point source with rate of introduction of mass $q(t)$ (kg/s) located at $\vec{y}$, equation (3.39) can be rewritten by using the three-dimensional delta function (see Section 3.2.3)

$$\frac{\partial^2 \rho'}{\partial t^2}-c_0^2\frac{\partial^2 \rho'}{\partial x_i{}^2}=\dot{q}(t)\cdot\delta(\vec{x}-\vec{y}). \tag{3.40}$$

$\dot{q}(t)$ has the dimension kg/s^2. As stated above, the solution of equation (3.40) for an unbounded medium in three dimensions can be found by the methods of Green's functions [168], [79] and is given by

$$4\pi\cdot p'(\vec{x},t)=\frac{\dot{q}(t-r/c_0)}{r}=\frac{\dot{q}(\tau)}{r}=\frac{[\dot{q}]}{r} \tag{3.41}$$

where r is the distance from the point source at $\vec{y}$ to a field point $\vec{x}$

$$r=|\vec{x}-\vec{y}|. \tag{3.42}$$

$\vec{x}$ is called the observer position or simply the observer. The acoustic pressure p' depends on the value of $\dot{q}$ at the *retarded time* $\tau = t - r/c_0$, i.e. at that time a wave had to leave the source to reach the observer at t. It is common to enclose terms which are taken at the retarded time τ in square brackets

$$g(t - r/c_0) = g(\tau) = [g]. \tag{3.43}$$

Since the retarded time τ represents the time when the signal was emitted, it is called the *emission time*. Accordingly, the time when the signal reaches the observer is called *the observer time*. The difference between emission and observer time is due to the finite speed of propagation.

In the frequency domain, the sound field radiated from a harmonic point source is given by

$$4\pi \cdot \tilde{p}'(\vec{x}, \omega) = \tilde{q}(\omega) \cdot \frac{e^{+ik_0 r}}{r}. \tag{3.44}$$

Note that this result follows directly from equations (3.34) and (3.35). For a continuous source field of density $Q(\vec{x}, t)$, the sound field is given by an integral over the region of flow containing the sources

$$4\pi \cdot p'(\vec{x}, t) = \int_V [\dot{Q}(\vec{y})] \frac{\mathrm{d}V(\vec{y})}{r}. \tag{3.45}$$

The sound field of a point source has monopole character as it is radial symmetric and depends only on the distance r from the source. Therefore, a point source is called a monopole with $\dot{q}$ being the monopole strength. $\dot{Q}$ is the monopole strength per unit volume. Figure 3.2 shows an instantaneous picture of the sound waves emitted by a monopole with harmonically fluctuating strength $\dot{q}$.

In practice, sound radiation with monopole character occurs with fluctuating sources of matter as in the case of a siren. Monopole sound is also radiated by moving volumes as in the case of a propeller or a fan. The most simple model for a monopole radiator is a small pulsating sphere.

3.4.2 Dipoles

If a fluctuating force field of density F_i (N/m^3) is distributed in part of the medium, the momentum equation (3.18) has to be rewritten

$$\frac{\partial}{\partial t}(\rho u_i) + \frac{\partial}{\partial x_j}(\rho u_i u_j) + \frac{\partial p}{\partial x_i} = F_i(\vec{x}, t). \tag{3.46}$$

Figure 3.2: Instantaneous picture of the sound waves emitted by a monopole; the waves spread uniformly and there is no directional dependence.

Performing the linearisation results again in an additional term on the right-hand side of the wave equation

$$\frac{\partial^2 \rho'}{\partial t^2} - c_0^2 \frac{\partial^2 \rho'}{\partial x_i^2} = -\frac{\partial F_i(\vec{x},t)}{\partial x_i} = -\frac{\partial F_1}{\partial x_1} - \frac{\partial F_2}{\partial x_2} - \frac{\partial F_3}{\partial x_3} \tag{3.47}$$

which has the dimension N/m^4. This term is the divergence of F_i and acts as a forcing term to the homogenous wave equation. In the following, the force field is to be non-zero only at the point $\vec{y}$. The total force exerted on this point has the magnitude f_i (N). As for the monopole above, the force density F_1 can be expressed by using the three-dimensional delta function (the following is analogous to the x_2- and x_3-direction)

$$F_1 = f_1 \cdot \delta(\vec{x} - \vec{y}) = f_1 \cdot \delta(x_1 - y_1)\, \delta(x_2 - y_2)\, \delta(x_3 - y_3)\,. \tag{3.48}$$

The delta function in the x_1-direction can be expressed by two Heaviside functions $H(x)$

$$F_1 = f_1 \cdot \lim_{\varepsilon_1 \to 0} \left\{ \frac{H\left(x_1 - \left\{y_1 - \frac{\varepsilon_1}{2}\right\}\right) - H\left(x_1 - \left\{y_1 + \frac{\varepsilon_1}{2}\right\}\right)}{\varepsilon_1} \right\} \cdot \delta(x_2 - y_2)\, \delta(x_3 - y_3)\,. \tag{3.49}$$

Performing the derivative $-\partial F_1/\partial x_1$ yields

$$-\frac{\partial F_1}{\partial x_1} = -f_1 \cdot \lim_{\varepsilon_1 \to 0} \left\{ \frac{\delta\left(x_1 - \left\{y_1 - \frac{\varepsilon_1}{2}\right\}\right) - \delta\left(x_1 - \left\{y_1 + \frac{\varepsilon_1}{2}\right\}\right)}{\varepsilon_1} \right\} . \qquad (3.50)$$

$$\cdot\, \delta(x_2 - y_2)\, \delta(x_3 - y_3)$$

Note that the derivative in the x_1-direction has transformed the pair of Heaviside functions into a pair of delta functions (see Section 3.2.3). Combining the delta functions yields

$$-\frac{\partial F_1}{\partial x_1} = \lim_{\varepsilon_1 \to 0} \left\{ -\frac{f_1}{\varepsilon_1} \cdot \delta\left(\vec{x} - \vec{y}_{-\varepsilon_1}\right) + \frac{f_1}{\varepsilon_1} \cdot \delta\left(\vec{x} - \vec{y}_{+\varepsilon_1}\right) \right\} \qquad (3.51)$$

with

$$\vec{y}_{-\varepsilon_1} = \left(y_1 - \frac{\varepsilon_1}{2} \,\middle|\, y_2 \,\middle|\, y_3 \right) \qquad \vec{y}_{+\varepsilon_1} = \left(y_1 + \frac{\varepsilon_1}{2} \,\middle|\, y_2 \,\middle|\, y_3 \right) . \qquad (3.52)$$

Comparison with equation (3.40) shows that the sound field, due to a concentrated force of magnitude f_1 in the x_1-direction, can be created with two point sources: one source of strength $+f_1/\varepsilon_1$ is located at $\vec{y}_{+\varepsilon_1}$, the second source of strength $-f_1/\varepsilon_1$ is located at $\vec{y}_{-\varepsilon_1}$.

Although equation (3.51) postulates that ε_1 approaches zero, ε_1 is sufficiently small if (1) the distance from a field point to $\vec{y}$ is large compared to ε_1 and (2) ε_1 is small compared to the wavelength of the radiated sound λ. Then, the field depends only on the *product* of the monopole-strength f_1/ε_1 and the separation ε_1. Thus, the two monopoles constitute a *dipole* of *dipole strength* f_1 with axis in the x_1-direction. f_1 is also called the *moment of the dipole.*

Equation (3.47) implies two additional dipoles of strength f_2 and f_3 aligned with the x_2- and x_3-axis respectively. Together, they form a dipole of strength $\vec{f}$ whose *dipole axis* is aligned with the force vector as depicted in Figure 3.3. In the case of a continuous force field, the force density $\vec{F}$ is the *dipole strength per unit volume*. In both cases the dipole strength is a vector.

The sound field due to a concentrated point force for an unbounded medium can be calculated by applying the Green's function for the wave equation

  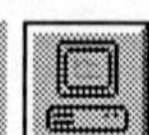 

$$4\pi \cdot p'(\vec{x},t) = -\frac{\partial}{\partial x_i}\left(\frac{[f_i]}{r}\right). \tag{3.53}$$

The corresponding solution in the frequency domain is given by

$$4\pi \cdot \tilde{p}'(\vec{x},\omega) = -\tilde{f}_i(\omega)\frac{\partial}{\partial x_i}\left(\frac{e^{ik_0 r}}{r}\right) \tag{3.54}$$

The characteristics of dipole radiation become clearer when the space derivatives in equations (3.53) and (3.54) are evaluated

$$4\pi \cdot p'(\vec{x},t) = \frac{x_i - y_i}{r}\left\{\frac{1}{rc_0}\left[\frac{\partial f_i}{\partial t}\right] + \frac{1}{r^2}[f_i]\right\} \tag{3.55}$$

$$4\pi \cdot \tilde{p}'(\vec{x},\omega) = \tilde{f}_i(\omega)\frac{x_i - y_i}{r}\left(-\frac{ik_0}{r} + \frac{1}{r^2}\right)e^{ik_0 r}. \tag{3.56}$$

Equations (3.55) and (3.56) show that the dipole field has two regions: a near field where the pressure fluctuations are dominated by the second term which falls off by $1/r^2$ and a far field that falls off by $1/r$. Thus, the difference between the pressure fluctuations in the strong near field and the weaker far field are much more pronounced than in the case of monopole radiation. Therefore, dipole radiation is said to be inefficient compared to monopole radiation because less energy from the relatively intense near field is radiated away as sound. This is due to the reason that the fluid can move *back and forth* between the two sources and *compression* – which leads to acoustic radiation – is at least partially avoided.

If the near field term is neglected (for $k_0 r >> 1$) and a field position is defined by the distance r and the angle Φ (see Figure 3.4) the sound field radiated from a point force can be expressed in the frequency domain

$$4\pi \cdot \tilde{p}'(\vec{x},\omega) = -cos(\Phi) ik_0 \tilde{f}(\omega) \cdot \frac{e^{ik_0 r}}{r} \tag{3.57}$$

Equation (3.57) clearly displays the most prominent characteristic of dipole radiation (see Figure 3.4): the space derivative in the forcing term introduces a directional dependence into the sound field with a maximum radiation in the direction of the dipole axis ($\Phi = 0, \pi$) and zero radiation perpendicular to it ($\Phi = \pi/2, 3\pi/2$). The latter occurs because the two monopole fields cancel each other totally if the distance of the observer is the same.

For a continuous force field of density F_i the sound field is given by an integral over the region of flow containing the force field

$$4\pi \cdot p'(\vec{x},t) = -\frac{\partial}{\partial x_i} \int_V [F_i] \frac{\mathrm{d}V(\vec{y})}{r} \quad . \tag{3.58}$$

If the space derivative is evaluated and the near field neglected, equation (3.58) reduces to

$$4\pi \cdot p'(\vec{x},t) = \frac{1}{c_0} \int_V \frac{x_i - y_i}{r} \left[\frac{\partial F_i}{\partial t} \right] \frac{\mathrm{d}V(\vec{y})}{r} \quad . \tag{3.59}$$

Figure 3.5 shows an instantaneous picture of the sound waves emitted by a dipole with harmonically fluctuating dipole strength f .

In practice, sound with dipole character occurs when fluctuating and/or moving forces are exerted on the fluid, for example, by moving surfaces as in the case of a propeller or a fan. The most simple model for a dipole radiator is a small oscillating sphere.

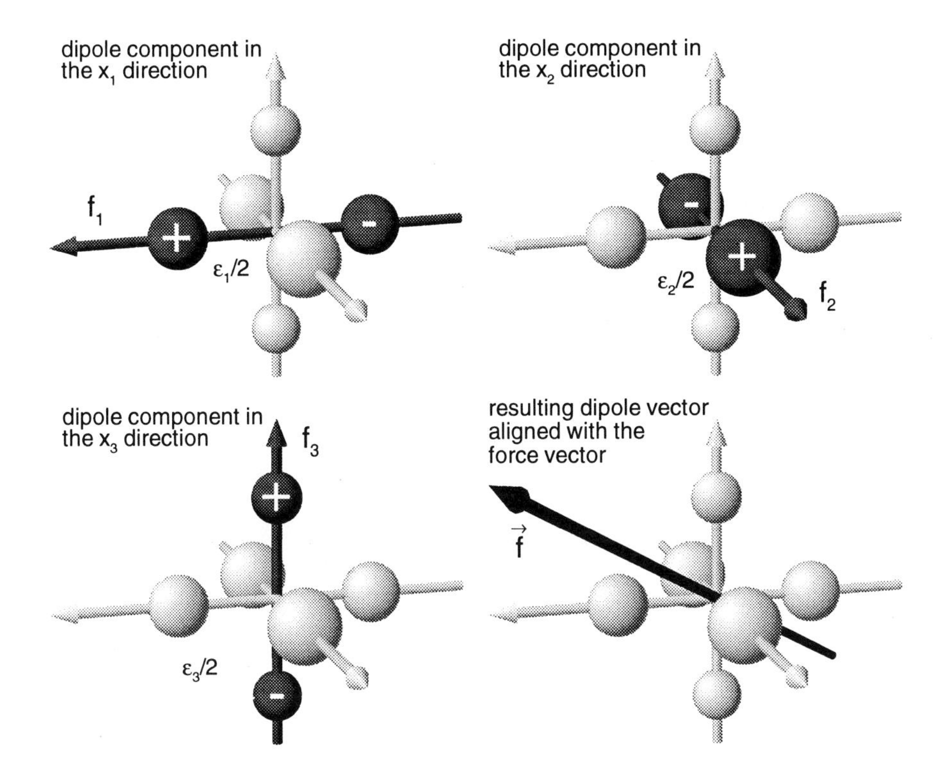

Figure 3.3: Each component f_i of the force vector acts as a dipole aligned with the x_i-axis. Together, they form a dipole of strength $\vec{f}$ aligned with the force vector.

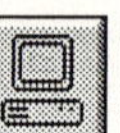

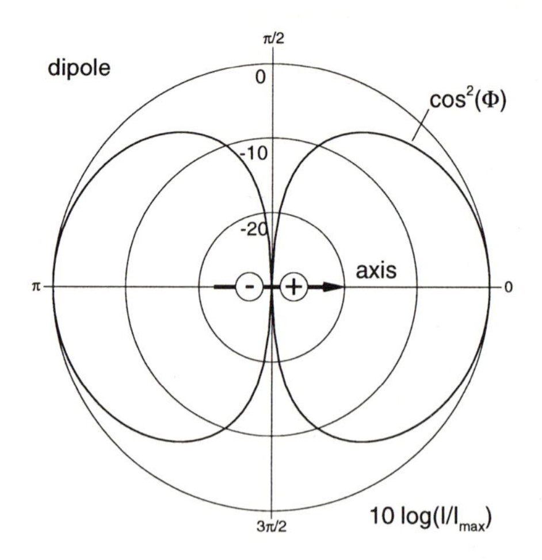

Figure 3.4: Directivity pattern of a dipole.

Figure 3.5: Instantaneous picture of the sound waves emitted by a dipole; the maximum radiation occurs in the direction of the dipole axis whereas no sound is radiated perpendicular to it.

3.5 Lighthill's Acoustic Analogy

The starting point is the continuity equation and the momentum equation in the ith direction for viscous flow (Navier-Stokes equation) [22], [139], [141]. A bounded region in space is occupied by a time-fluctuating turbulent fluid motion. Outside this region the flow is assumed to be at rest with uniform pressure and density p_0 and ρ_0, respectively. The fluid is to be free of sources of matter or external forces

$$\frac{\partial \rho}{\partial t}+\frac{\partial}{\partial x_i}(\rho u_i)=0 \tag{3.60}$$

$$\begin{gathered}\frac{\partial}{\partial t}(\rho u_i)+\frac{\partial}{\partial x_j}\left(\rho u_i u_j+p_{ij}\right)=0\\ p_{ij}=p\delta_{ij}+\mu\left\{-\frac{\partial u_i}{\partial x_j}-\frac{\partial u_j}{\partial x_i}+\frac{2}{3}\left(\frac{\partial u_k}{\partial x_k}\right)\delta_{ij}\right\}.\end{gathered} \tag{3.61}$$

Here p_{ij} denotes the compressive stress tensor, δ_{ij} is the Kronecker delta, and μ the viscosity. As in Section 3.3, the time derivative of the continuity equation and the divergence of the momentum equation are taken. Both equations are combined in order to remove the momentum density ρu_i

$$\frac{\partial^2 \rho}{\partial t^2}=\frac{\partial^2}{\partial x_i \partial x_j}\left(\rho u_i u_j+p_{ij}\right). \tag{3.62}$$

Now, the term $c_0^2\,\partial^2\rho/\partial x_i^2$ is subtracted from both sides, the undisturbed pressure and density p_0 and ρ_0 is inserted and the terms on the right-hand side are rearranged to a tensor T_{ij} which is called the Lighthill tensor:

$$\boxed{\begin{gathered}\frac{\partial^2 \rho'}{\partial t^2}-c_0^2\frac{\partial^2\rho'}{\partial x_i^2}=\frac{\partial^2 T_{ij}}{\partial x_i \partial x_j}\\ T_{ij}=\rho u_i u_j+\left(p-p_0-c_0^2(\rho-\rho_0)\right)\delta_{ij}+\mu\left[-\frac{\partial u_i}{\partial x_j}-\frac{\partial u_j}{\partial x_i}+\frac{2}{3}\left(\frac{\partial u_k}{\partial x_k}\right)\delta_{ij}\right]\end{gathered}}. \tag{3.63}$$

$\rho u_i u_j$ denotes the convection of the momentum component ρu_i by the velocity component u_j, $(p-p_0 - c_0^2(\rho-\rho_0))$ is the departure from a state where all fluctuations are isentropic, and the last term is the transport of momentum due to viscous stresses. T_{ij} has the dimension kg/(ms^2). If all fluctuations are

isentropic, equation (3.25) is valid throughout the fluid and $p-p_0$ and $c_0^2(\rho-\rho_0)$ cancel each other. Furthermore, in turbulent flows the direct convection of momentum $\rho u_i u_j$ (fluctuating Reynolds stresses) is much larger than the transport of momentum due to viscous stresses. Equation (3.63) then reduces to

$$\frac{\partial^2 \rho'}{\partial t^2} - c_0^2 \frac{\partial^2 \rho'}{\partial x_i^2} = \frac{\partial^2}{\partial x_i \partial x_j}\left(\rho u_i u_j\right) \tag{3.64}$$

with the Lighthill tensor T_{ij} taking the simpler form

$$T_{ij} = \rho u_i u_j = \rho \begin{pmatrix} u_1^2 & u_1 u_2 & u_1 u_3 \\ u_2 u_1 & u_2^2 & u_2 u_3 \\ u_3 u_1 & u_3 u_2 & u_3^2 \end{pmatrix}. \tag{3.65}$$

Equation (3.64) has been derived by simply rearranging the continuity and momentum equation in a way that makes them look like an inhomogenous wave equation. The non-linear term $\rho u_i u_j$ in the momentum equation appears as a forcing term on the right-hand side of the wave equation. This *convection of momentum* which is dominated by the turbulent motion of the fluid acts as a source to a linearly responding medium outside the region of turbulent flow. Physically the Lighthill tensor is a *system of stresses* applied to a fluid element. Remember that $\rho u_i u_j$ is exactly the term that has been neglected in the derivation of the homogeneous wave equation in Section 3.2. Outside the region of turbulent flow, equation (3.64) reduces to the homogenous wave equation (3.27).

The important conclusion is that Lighthill's theory allows to compute the sound field radiated by a bounded region of fluctuating (turbulent) flow by solving an *analogous* problem of forced oscillation, *provided that the flow is known*. This is called Lighthill's acoustic analogy.

The forcing term on the right-hand side of the wave equation includes a double space derivative. A forcing term without space derivative (see equation (3.39)) resulted in an omnidirectional monopole field. A forcing term with one space derivative (see equation (3.47)) resulted in a dipole field with a two-lobed directivity distribution. The following will show that a double space derivative implies a quadrupole field with a four-lobed and a two-lobed directivity distribution depending on whether the space derivatives are performed in different and the same direction, respectively.

3.5.1 Quadrupoles

The inhomogenous wave equation with the Lighthill tensor as forcing term is

$$\frac{\partial^2 \rho'}{\partial t^2} - c_0^2 \frac{\partial^2 \rho'}{\partial x_i^2} = \frac{\partial^2 T_{ij}}{\partial x_i \partial x_j} = +\frac{\partial^2 T_{11}}{\partial x_1^2} + \frac{\partial^2 T_{12}}{\partial x_1 \partial x_2} + \frac{\partial^2 T_{13}}{\partial x_1 \partial x_3}$$
$$+\frac{\partial^2 T_{21}}{\partial x_2 \partial x_1} + \frac{\partial^2 T_{22}}{\partial x_2^2} + \frac{\partial^2 T_{23}}{\partial x_2 \partial x_3} . \qquad (3.66)$$
$$+\frac{\partial^2 T_{31}}{\partial x_3 \partial x_1} + \frac{\partial^2 T_{32}}{\partial x_3 \partial x_2} + \frac{\partial^2 T_{33}}{\partial x_3^2}$$

In the following, the characteristics of quadrupole radiation will be examined for the forcing term $\partial^2 T_{12}/\partial x_1 \partial x_2$. T_{12} is to be non-zero only within a vanishingly small cube (see the top of Figure 3.6). The *concentrated* strength, i.e. the integral of T_{12} over the small cube, is t_{12} (kgm^2/s^2). Since the volume of the cube tends to zero, T_{12} must approach infinity. As above, this can be expressed by using the three-dimensional delta function

$$T_{12} = t_{12} \cdot \delta(\vec{x} - \vec{y}) = t_{12} \cdot \delta(x_1 - y_1)\, \delta(x_2 - y_2)\, \delta(x_3 - y_3) . \qquad (3.67)$$

The delta function in the x_1- and x_2-direction can be expressed by two Heaviside functions $H(x)$

$$T_{12} = t_{12} \cdot \lim_{\substack{\varepsilon_1 \to 0 \\ \varepsilon_2 \to 0}} \left\{ \frac{H\left(x_1 - \left\{y_1 - \frac{\varepsilon_1}{2}\right\}\right) - H\left(x_1 - \left\{y_1 + \frac{\varepsilon_1}{2}\right\}\right)}{\varepsilon_1} \cdot \frac{H\left(x_2 - \left\{y_2 - \frac{\varepsilon_2}{2}\right\}\right) - H\left(x_2 - \left\{y_2 + \frac{\varepsilon_2}{2}\right\}\right)}{\varepsilon_2} \right\} \cdot \delta(x_3 - y_3). \qquad (3.68)$$

Performing the derivatives in the x_1- and x_2-direction yields.

$$\frac{\partial^2 T_{12}}{\partial x_2 \partial x_1} = t_{12} \cdot \lim_{\substack{\varepsilon_1 \to 0 \\ \varepsilon_2 \to 0}} \left\{ \frac{\delta\left(x_1 - \left\{y_1 - \frac{\varepsilon_1}{2}\right\}\right) - \delta\left(x_1 - \left\{y_1 + \frac{\varepsilon_1}{2}\right\}\right)}{\varepsilon_1} \cdot \frac{\delta\left(x_2 - \left\{y_2 - \frac{\varepsilon_2}{2}\right\}\right) - \delta\left(x_2 - \left\{y_2 + \frac{\varepsilon_2}{2}\right\}\right)}{\varepsilon_2} \right\} \cdot \delta(x_3 - y_3). \qquad (3.69)$$

Note that the derivatives have transformed the pair of Heaviside functions into a pair of delta functions. Combining the delta functions yields the following formula

$$\frac{\partial^2 T_{12}}{\partial x_2 \partial x_1} = \lim_{\substack{\varepsilon_1 \to 0 \\ \varepsilon_2 \to 0}} \left\{ \begin{array}{l} +\dfrac{t_{12}}{\varepsilon_1 \varepsilon_2} \cdot \delta\left(\vec{x} - \vec{y}_{-\varepsilon_1 - \varepsilon_2}\right) - \dfrac{t_{12}}{\varepsilon_1 \varepsilon_2} \cdot \delta\left(\vec{x} - \vec{y}_{-\varepsilon_1 + \varepsilon_2}\right) \\ -\dfrac{t_{12}}{\varepsilon_1 \varepsilon_2} \cdot \delta\left(\vec{x} - \vec{y}_{+\varepsilon_1 - \varepsilon_2}\right) + \dfrac{t_{12}}{\varepsilon_1 \varepsilon_2} \cdot \delta\left(\vec{x} - \vec{y}_{+\varepsilon_1 + \varepsilon_2}\right) \end{array} \right\} \tag{3.70}$$

with

$$\begin{array}{ll} \vec{y}_{-\varepsilon_1 - \varepsilon_2} = \left(y_1 - \dfrac{\varepsilon_1}{2} \;\middle|\; y_2 - \dfrac{\varepsilon_2}{2} \;\middle|\; y_3 \right) & \vec{y}_{-\varepsilon_1 + \varepsilon_2} = \left(y_1 - \dfrac{\varepsilon_1}{2} \;\middle|\; y_2 + \dfrac{\varepsilon_2}{2} \;\middle|\; y_3 \right) \\ \vec{y}_{+\varepsilon_1 - \varepsilon_2} = \left(y_1 + \dfrac{\varepsilon_1}{2} \;\middle|\; y_2 - \dfrac{\varepsilon_2}{2} \;\middle|\; y_3 \right) & \vec{y}_{+\varepsilon_1 + \varepsilon_2} = \left(y_1 + \dfrac{\varepsilon_1}{2} \;\middle|\; y_2 + \dfrac{\varepsilon_2}{2} \;\middle|\; y_3 \right) \end{array}. \tag{3.71}$$

Hence, the radiation field of a small cube suffering from a *concentrated* stress t_{12} can be expressed as a superposition of the fields of four sources, two with positive, two with negative sign which form a cross as depicted in Figure 3.6. The physical meaning of equation (3.70) becomes clear if the concentrated quantity t_{12} is replaced by $T_{12}\varepsilon_1\varepsilon_2\varepsilon_3$, i.e. T_{12} times the volume of the small cube. Equation (3.70) can be rearranged as

$$\frac{\partial^2 T_{12}}{\partial x_2 \partial x_1} = \lim_{\substack{\varepsilon_1 \to 0 \\ \varepsilon_2 \to 0}} \left\{ \begin{array}{l} -\left\{ -\dfrac{T_{12}\varepsilon_1\varepsilon_3}{\varepsilon_1} \cdot \delta\left(\vec{x} - \vec{y}_{-\varepsilon_1 - \varepsilon_2}\right) + \dfrac{T_{12}\varepsilon_1\varepsilon_3}{\varepsilon_1} \cdot \delta\left(\vec{x} - \vec{y}_{+\varepsilon_1 - \varepsilon_2}\right) \right\} \\ +\left\{ -\dfrac{T_{12}\varepsilon_1\varepsilon_3}{\varepsilon_1} \cdot \delta\left(\vec{x} - \vec{y}_{-\varepsilon_1 + \varepsilon_2}\right) + \dfrac{T_{12}\varepsilon_1\varepsilon_3}{\varepsilon_1} \cdot \delta\left(\vec{x} - \vec{y}_{+\varepsilon_1 + \varepsilon_2}\right) \right\} \end{array} \right\}. \tag{3.72}$$

Comparison with equation (3.51) shows the analogy to the forcing term for dipole radiation: the sources in the brackets above constitute a dipole of strength $-T_{12}\varepsilon_1\varepsilon_3$ aligned with the x_1-axis. The sources in the brackets below constitute a parallel dipole of strength $+T_{12}\varepsilon_1\varepsilon_3$. The dipoles are separated in the x_2-direction. Since $T_{12}\varepsilon_1\varepsilon_3$ is exactly the force which acts onto a surface of area $\varepsilon_1\varepsilon_3$ that suffers from a stress T_{12}, the physical meaning of radiation due to turbulent fluid motion becomes clear: a small fluid element suffering from a fluctuating Reynolds stress T_{12} radiates sound like two dipoles acting onto the opposite surfaces of the fluid element, the dipole strength being exactly the force created by the stress at the surface. Since the *two dipoles* can be expressed as a pair of monopoles, the sound field is made up by *four*

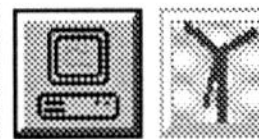

monopoles. Therefore, radiation due to turbulent fluid motion is called *quadrupole* radiation.

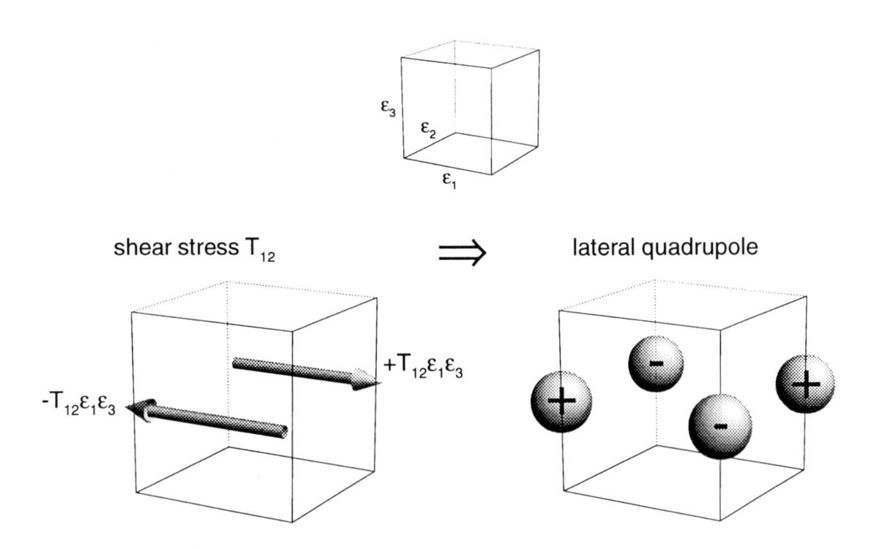

Figure 3.6: Lateral quadrupole: the component T_{12} of the Lighthill tensor radiates sound like two dipoles or four monopoles which form a cross.

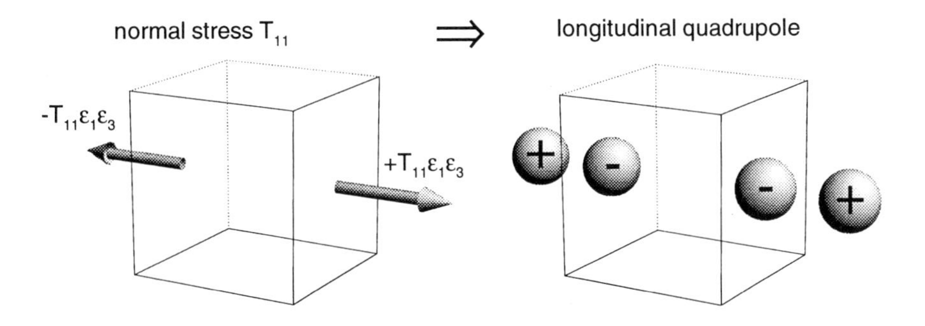

Figure 3.7: Longitudinal quadrupole: the component T_{11} of the Lighthill tensor radiates sound like two dipoles or four monopoles in a line.

Although equation (3.72) postulates that ε_1 and ε_2 equally tend to zero, the combined field of these monopoles depends only on the *product* of the monopole strength $T_{12}\varepsilon_3$ and $\varepsilon_1\varepsilon_2$, if (1) the distance from a field point to $\vec{y}$ is large compared to $\varepsilon_1,\varepsilon_2$, and (2) $\varepsilon_1,\varepsilon_2$ are small compared to the wavelength λ. Then, the four sources will constitute a quadrupole of strength $t_{12} = T_{12}\varepsilon_1\varepsilon_2\varepsilon_3$.

Accordingly, there are quadrupoles for the other eight components of the Lighthill tensor. Depending on whether the space derivatives are performed both in the same direction or not, the result is a longitudinal or a lateral quadrupole, respectively. In the first case the two dipoles are aligned and the four sources are arranged in one line (see Figure 3.7), in the second case the two dipoles are parallel and the four sources form a cross (see Figure 3.6).

The sound field due to a point quadrupole of strength t_{ij} for an unbounded medium can be calculated by applying the Green's function for the wave equation

$$4\pi \cdot p'(\vec{x},t) = \frac{\partial^2}{\partial x_i \partial x_j}\left(\frac{[t_{ij}]}{r}\right). \tag{3.73}$$

The solution in the frequency domain is given by

$$4\pi \cdot \tilde{p}'(\vec{x},\omega) = \tilde{t}_{ij}(\omega)\frac{\partial^2}{\partial x_i \partial x_j}\left(\frac{e^{ik_0 r}}{r}\right). \tag{3.74}$$

The characteristics of quadrupole radiation become clearer when the space derivatives in equations (3.73) and (3.74) are evaluated

$$4\pi \cdot \tilde{p}'(\vec{x},\omega) = \tilde{t}_{ij}(\omega)\frac{x_i - y_i}{r}\frac{x_j - y_j}{r}\left(-\frac{k_0^2}{r} - \frac{3ik_0}{r^2} + \frac{3}{r^3}\right)e^{ik_0 r} \tag{3.75}$$

$$4\pi \cdot p'(\vec{x},t) = \frac{x_i - y_i}{r}\frac{x_j - y_j}{r}\left(\frac{1}{rc_0^2}\left[\frac{\partial^2 t_{ij}}{\partial t^2}\right] + \frac{3}{r^2 c_0}\left[\frac{\partial t_{ij}}{\partial t}\right] + \frac{3}{r^3}[t_{ij}]\right). \tag{3.76}$$

Equations (3.75) and (3.76) show that the quadrupole field has now three regions: a near field where the pressure fluctuations are dominated by the last term which falls off by $1/r^3$ (remember $1/r^2$ for dipole radiation), a far field that falls off by $1/r$, and a region in between which falls of by $1/r^2$. Thus, the difference between the pressure fluctuations in the strong near field and the weaker far field are even more pronounced than in the case of dipole radiation. Therefore, quadrupole radiation is even less efficient than dipole radiation which in turn is less efficient than monopole radiation. This is again due to the fact that a relatively small amount of energy from the intense near field is radiated away as sound, because the fluid can move *back and forth* between the four sources and *compression* – which leads to acoustic radiation – is at least partially avoided.

If the near-field terms are neglected (for $k_0 r >> 1$) and a field position is defined by the distance r and the angle Φ (see Figure 3.8) the sound field radiated from a longitudinal quadrupole is given by

$$4\pi \cdot \tilde{p}'(\vec{x},\omega) = -cos^2(\Phi) k_0^2 \tilde{t}_{ij}(\omega) \cdot \frac{e^{+ik_0 r}}{r}. \tag{3.77}$$

The sound field radiated from a lateral quadrupole is given by

$$4\pi \cdot \tilde{p}'(\vec{x},\omega) = -sin(2\Phi) k_0^2 \tilde{t}_{ij}(\omega) \cdot \frac{e^{+ik_0 r}}{r}. \tag{3.78}$$

Equations (3.77) and (3.78) clearly display the most prominent characteristic of quadrupole radiation: for a lateral quadrupole the double space derivative in the forcing term introduces a four-lobed directional dependence into the sound field as depicted in Figure 3.8. The field of a longitudinal quadrupole looks like a dipole field. However, in contrast to dipole radiation the fluctuations in the two lobes are *in phase*. The occurrence of these directivity patterns is again due to the cancellation of the fields of the four monopoles if the distance to a field point is equal.

For a continuous distribution of fluctuating Reynolds stresses T_{ij} the sound field is given by an integral over the region of flow which contains the stresses

$$4\pi \cdot p'(\vec{x},t) = \frac{\partial^2}{\partial x_i \partial x_j} \int_V \left[T_{ij}\right] \frac{\mathrm{d}V(\vec{y})}{r}. \tag{3.79}$$

If the space derivatives are evaluated and the near-field terms are neglected, equation (3.79) reduces to

$$4\pi \cdot p'(\vec{x},t) = \frac{1}{c_0^2} \int_V \frac{x_i - y_i}{r} \frac{x_j - y_j}{r} \left[\frac{\partial^2 T_{ij}}{\partial t^2}\right] \frac{\mathrm{d}V(\vec{y})}{r}. \tag{3.80}$$

Figure 3.9 and Figure 3.10 show an instantaneous picture of the sound waves emitted by a lateral and longitudinal quadrupole of harmonically fluctuating quadrupole strength t_{ij}, respectively. In practice, sound with quadrupole character occurs in all turbulent flows. The most simple model for a quadrupole radiator is a small deforming sphere. Table 3.1 summarizes the properties of the three elementary solutions of the wave equation, namely monopoles, dipoles, and quadrupoles.

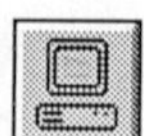

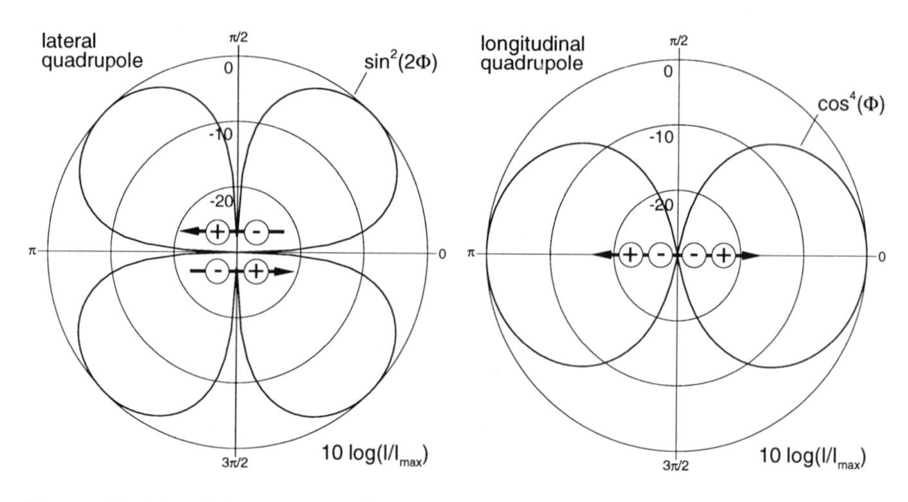

Figure 3.8: Directivity patterns of lateral and longitudinal quadrupoles.

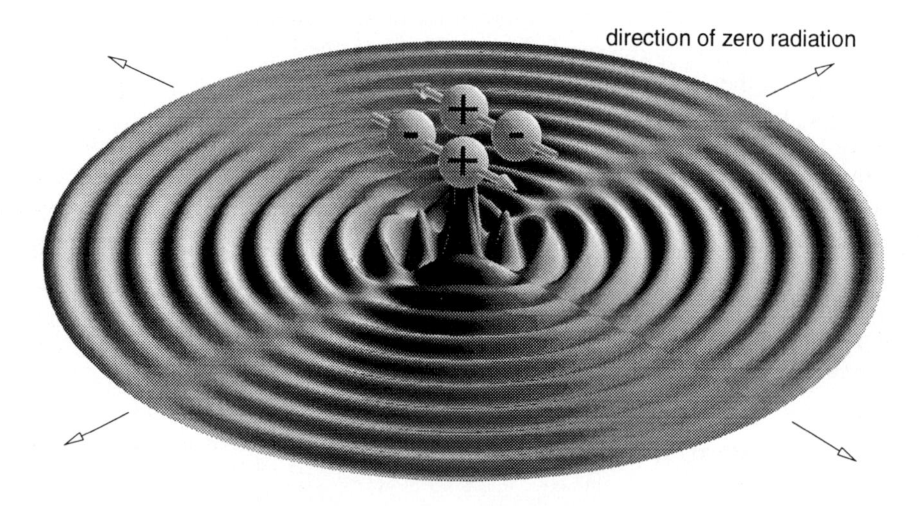

Figure 3.9: Instantaneous picture of the sound field emitted by a lateral quadrupole; no sound is radiated in the direction of the dipoles and perpendicular to it, whereas maximum radiation occurs in between.

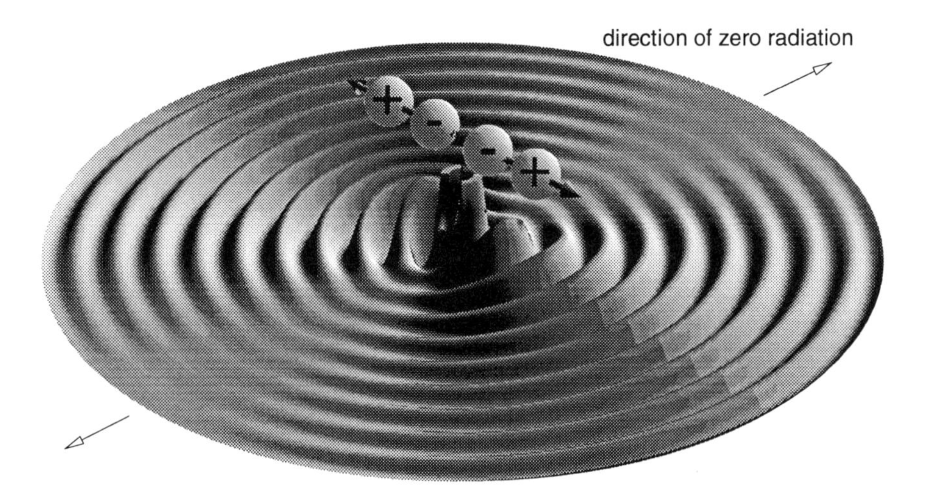

Figure 3.10: Instantaneous picture of the sound field emitted by a longitudinal quadrupole; maximum radiation occurs in the direction of the two dipoles whereas no sound is radiated perpendicular to it.

Table 3.1: Properties of elementary solutions of the wave equation.

	Monopole	*Dipole*	*Quadrupole*
Physical mechanism	Fluctuating injection of matter	Fluctuating injection of momentum, unsteady force acting onto the fluid	System of normal and shear stresses, fluctuating Reynolds stresses
Strength	Rate of change of mass inflow /outflow, scalar quantity	Force acting onto the fluid, vector quantity	Lighthill tensor T_{ij}, tensor quantity
Directivity	Omnidirectional	Two lobes	Four lobes for lateral, two lobes for longitudinal quadrupole
Practical meaning	Siren, moving volume	Rotating, fluctuating forces on blades, propellers, fans	Turbulent flows

3.5.2 Powell's Theory of Vortex Sound

Lighthill expressed the sound field which is radiated from a turbulent fluid motion as a volume integral of the Lighthill tensor T_{ij} which in turn is dominated by the fluctuating Reynolds stresses $\rho u_i u_j$. On the other hand, Powell investigated which characteristics of the turbulent flow or eddy motion are responsible for sound generation. He found that the formation and motion

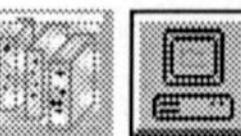

of vortices or vorticity is the fundamental noise-producing mechanism. A detailed description is given in [22], [179].

The starting point in Powell's analysis is again the equation of continuity and momentum. However, the second has been expressed in terms of the vorticity vector $\vec{\omega}$

$$\rho \frac{\partial \vec{u}}{\partial t} + \rho(\vec{\omega} \times \vec{u}) + \nabla\left(\frac{\rho u^2}{2}\right) = -\nabla p . \tag{3.81}$$

Both equations can be combined as in Section 3.5 in order to obtain the inhomogenous wave equation for the density variation with Powell's source term on the right hand side

$$\frac{1}{c_0^2} \frac{\partial^2 p'}{\partial t^2} - \nabla^2 p' = \nabla \cdot \left\{ \rho(\vec{\omega} \times \vec{u}) + \nabla\left(p - c_0^2 \rho + \frac{\rho u^2}{2} \right) \right\} . \tag{3.82}$$

In equation (3.82) all terms are neglected which are of second order for $|\vec{u}| << c_0$. It is important to note that the acoustic analogies by Powell [179] and Lighthill [139] are physically equivalent [22]. Another formulation of the acoustic analogy has been proposed by Howe [105].

3.5.3 Acoustical Compactness

An important concept in aeroacoustics is the acoustical compactness of a region that contains aeroacoustic sources. A source region of typical dimension l is defined to be acoustically compact if l is small compared to the wavelength λ of the radiated sound. Mathematically the condition of acoustical compactness can be written as

$$\frac{l}{\lambda} << 1 \qquad \text{or} \qquad k_0 l << 1 . \tag{3.83}$$

If the condition is fulfilled, differences in retarded time at different positions of the source region can be neglected. This will considerably simplify the evaluation of the volume integrals in equations (3.45), (3.59), (3.80).

3.6 The Influence of Boundaries

Lighthill's theory of aerodynamic sound describes the sound field emitted by a bounded region of fluctuating flow by considering a corresponding fluctuating stress field T_{ij} which acts onto a linearly responding acoustic medium at rest [139]. The main limitation of Lighthill's equation (3.63) is that it is strictly valid only for an unbounded fluid. Therefore, its application is limited to problem like jet noise where solid surfaces do not play a major role [143]. For aerodynamic noise radiated from blades (airfoil self noise) the influence of boundaries cannot be neglected. Some mechanisms of noise generation are even dominated by the presence of a surface in the flow.

The first extension of Lighthill's theory was proposed by Curle [37]. He considered the influence of static surfaces. Ffowcs Williams and Hawkings [65] included the influence of arbitrarily moving surfaces. Their generalized approach will be outlined in this section (see also [64]).

3.6.1 Ffowcs Williams–Hawkings Equation

In [65] the fluid is partitioned into different regions by mathematical surfaces which represent the physical boundaries between a body and the surrounding fluid (see Figure 3.11). Outside the surfaces the flow is identical to the 'real' flow, inside the surfaces it can be specified arbitrarily. Since the interior flow normally does not match the exterior flow at the surfaces, additional source terms have to be placed on the boundaries in order to maintain these discontinuities between interior and exterior flow.

The surface S is defined by the equation $f(\vec{x},t)=0$. Note that the function $f(\vec{x},t)$ represents the *shape* of the surface and its *motion*. Ffowcs Williams and Hawkings start with the generalized equations of mass

$$\frac{\partial \rho}{\partial t}+\frac{\partial}{\partial x_i}(\rho u_i)=\rho_0 u_i \delta(f)\frac{\partial f}{\partial x_i} \tag{3.84}$$

and momentum

$$\frac{\partial}{\partial t}(\rho u_i)+\frac{\partial}{\partial x_j}(\rho u_i u_j+p_{ij})=p_{ij}\delta(f)\frac{\partial f}{\partial x_j}. \tag{3.85}$$

The difference to the equations (3.17) and (3.18) is the presence of mass and momentum source terms on the right-hand side of equations (3.84) and (3.85), respectively. These source terms are needed to maintain the unbounded fluid in its defined state. If there is only one mathematical region present, i.e. the

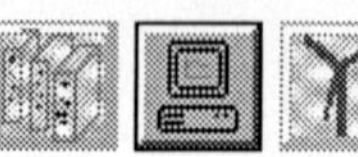

fluid region does not contain a solid surface, the generalized mass and momentum equation reduce to their ordinary form of an unbounded fluid (3.17), (3.18).

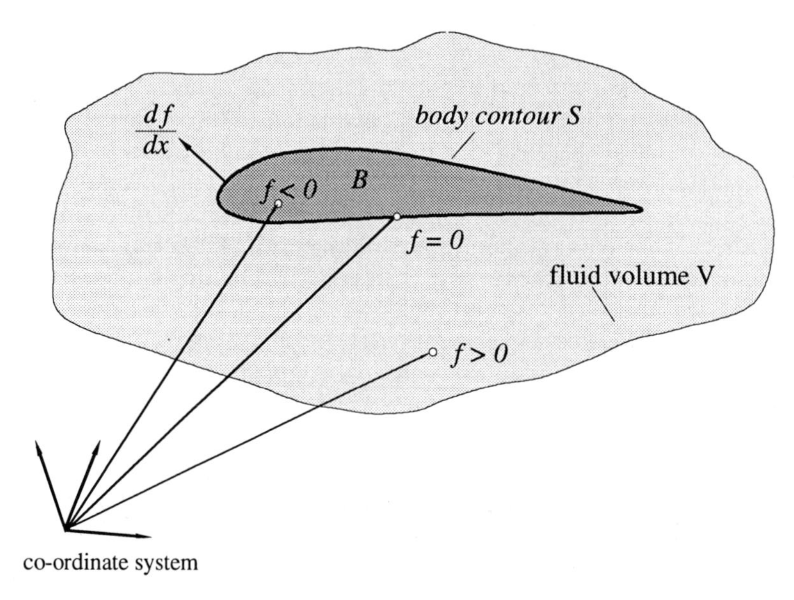

Figure 3.11: Body contour and fluid volume.

Following the same procedure as shown in Section 3.3 and 3.5, an inhomogenous wave equation is derived which is commonly labeled the Ffowcs Williams–Hawkings (FW–H) equation

$$\boxed{\frac{\partial^2 \rho'}{\partial t^2} - c_0^2 \frac{\partial^2 \rho'}{\partial x_i^2} = \frac{\partial^2 T_{ij}}{\partial x_i \partial x_j} - \frac{\partial}{\partial x_i}\left(p_{ij} \delta(f) \frac{\partial f}{\partial x_j} \right) + \frac{\partial}{\partial t}\left(\rho_0 u_i \delta(f) \frac{\partial f}{\partial x_i} \right)}. \quad 3.86)$$

The first term on the right-hand side of equation (3.86) is identical to Lighthill's term (see equation (3.63)). It represents the sound radiation due to fluctuating Reynolds stresses and is of *quadrupole* type. Its strength is given by the Lighthill tensor T_{ij} (see equation (3.63)).

Comparison with equation (3.47) shows that the second term is a *dipole* term. It is proportional to the stress tensor p_{ij} which contains the viscous stresses and the aerodynamic pressure. Since $\partial f / \partial x_j$ represents a vector which is outward normal to the surface $f(\vec{x}, t) = 0$, the dipole strength is given by the force vector which acts from the surface onto the fluid.

Comparison with equation (3.39) shows that the third term is a *monopole* term. It is proportional to the acceleration of the surface in the normal direction. If the surface is split into a set of elements, each surface element can be regarded as a small piston acting on the fluid with speed u_i. The monopole strength is given by the time derivative of the outward normal velocity u_n of the surface times the fluid density, i.e. the time derivative of a mass flow.

The important conclusion is that solid surfaces are acoustically equivalent to a surface distribution of monopoles and dipoles, with the respective strength being equal to the local acceleration of the surface and to the net force exerted onto the fluid.

Ffowcs Williams and Hawkings give different solutions for equation (3.86) which will not be treated in this chapter. Farassat [58] gives a general solution of the FW–H equation for the monopole and dipole terms. For the monopole term, which is called *thickness noise*, he obtains

$$4\pi \cdot p'(\vec{x},t) = \frac{\partial}{\partial t} \int_S \left[\frac{\rho_0 u_n}{r|1-M_r|} \right] \mathrm{d}S(\vec{y}) \tag{3.87}$$

where u_n denotes the surface velocity normal to the surface. For the dipole term, which is called *loading noise*, he obtains

$$4\pi \cdot p'(\vec{x},t) = -\frac{\partial}{\partial x_i} \int_S \left[\frac{p_{ij} n_j}{r|1-M_r|} \right] \mathrm{d}S(\vec{y}) \ . \tag{3.88}$$

Here, $p_{ij}n_j$ denotes the *force per unit area* acting from the surface onto the fluid. M_r is the relative Mach number of the source region $\vec{y}$, i.e. the component of the source velocity which is directed towards the observer at $\vec{x}$ divided by the speed of sound c_0. After expressing the derivatives in the frame of reference of the source and in terms of the source time τ, the compact solutions given in Section 5.2.1 can be derived [55], [58].

The FW-H equation is especially useful in cases where the aerodynamic forces of a moving body exerted on the surrounding fluid are known either by measurements or calculations and where effects of turbulence in the flow on noise are of minor importance. In Chapter 5, an application of the formulas to the case of low-frequency noise from wind turbine will be presented.

3.6.2 Helmholtz Integral Equation

The influence of boundaries can be investigated also in the frequency domain. Using the Green's function G, the inhomogenous Helmholtz equation (3.32) can be transformed into a boundary integral equation (Helmholtz integral

 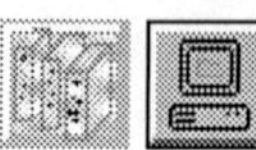 

equation) where the surface integrals have to be evaluated over all surfaces which are present in the flow [34]

$$\tilde{p}'(\vec{x},\omega) = \int_S \left\{ \frac{\partial \tilde{p}'(\vec{y},\omega)}{\partial n} G - \tilde{p}'(\vec{y},\omega) \frac{\partial G}{\partial n} \right\} \mathrm{d}S(\vec{y}) + \tilde{p}_i'(\vec{x},\omega) . \tag{3.89}$$

The first surface integral in equation (3.89) is of monopole type, whereas the second term includes a space derivative of the Green's function (fundamental solution) normal to the surface and is therefore a dipole term. With a linearised momentum equation it can be shown that there is a simple relationship between the acoustic pressure p' and the acoustic particle velocity u_i' in the time domain

$$\frac{\partial p'}{\partial x_i} = -\rho_0 \cdot \frac{\partial u_i'}{\partial t} \tag{3.90}$$

and in the frequency domain

$$\frac{\partial \tilde{p}'}{\partial x_i} = i\omega\rho_0 \cdot \tilde{u}_i' . \tag{3.91}$$

Comparison of equation (3.89) and (3.86) shows in combination with equation (3.91) that analogously to the time domain, the boundaries in the flow are represented by surface distributions of monopoles and dipoles where the strength is proportional to the acceleration of the surface in the normal direction and the forces which act onto the fluid, respectively. $\tilde{p}_i'$ denotes an incident wave which can, for example, result from a volume distribution of quadrupoles

$$\tilde{p}_i'(\vec{x},\omega) = \int_V \frac{\partial^2 \tilde{T}_{ij}(\vec{y},\omega)}{\partial y_i \partial y_j} G(\vec{x},\vec{y},\omega) \, \mathrm{d}V(\vec{y}) . \tag{3.92}$$

Formulations based on the Helmholtz equation are widely used due to the fact that they can be treated more easily mathematically than the wave equation. Especially the problem of reflection, diffraction, or scattering from bodies within the fluid is tackled in this way, as well as the sound generation from vibrating surfaces. Equation (3.89) can be solved numerically by discretizing the boundary into surface panels. A good introduction to the application of boundary element methods in acoustics can be found in [34] (see also [188], [11]).

3.7 Application of Aeroacoustic Theory

The previous sections have given an introduction to aeroacoustics and explained some important equations and concepts. This final section is thought as a link between these equations and the noise mechanisms and prediction formulas which are described in Chapter 4 and 5, respectively. It starts with an overview, which parts of aeroacoustic theory are suitable for the *actual calculation* of noise depending on the flow regimes and mechanisms of sound generation. The section continues with a set of basic cases which are relevant for wind turbine noise.

3.7.1 Approaches for Different Flow Regimes

Different approaches for the calculation of noise are needed for different flow regimes and noise mechanisms. In general, one can say that an actual *computation* of noise requires a knowledge of the details of the flow which cause the noise. For example, the steady, harmonic Gutin noise of a propeller is caused by the steady loads at the rotating blades [87]. These loads can be computed with good accuracy and a noise calculation is therefore possible without further empirical input. This type of noise can be termed as *deterministic* noise.

On the other hand, jet noise is caused by the turbulent flow which cannot be computed in all its details with presently available computers. Although Lighthill's equation gives an exact solution for the noise emitted by turbulent flows, a direct computation of the noise is therefore not possible because the details of the flow which determine the strength of the Lighthill tensor are not known. Since noise from wind turbines is caused by turbulence and its interaction with the blade surface, a noise prediction must contain a considerable amount of empirical input. This type of noise can be termed as *non-deterministic* (stochastic) noise. In the following, the different approaches will be outlined. Distinction will be made between low- and high- Mach number flows and between deterministic and non-deterministic noise.

Deterministic Noise, Subsonic Flows

Typical examples: Harmonic propeller or fan noise, low frequency noise from wind turbines.

Mechanism: The displacement of the air by the motion of the body gives rise to ⇒ **monopole sound**.
Forces act from the body onto the fluid ⇒ **dipole sound**.

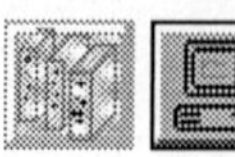

The loads at the blades can be computed using standard techniques (blade-element-momentum or vortex-lattice methods). These loads constitute source terms for the Ffowcs Williams–Hawkings equation. A solution of the FW–H equation is used to compute the sound radiation at arbitrary observer locations in the time domain [198], [57]. The non-linear terms which cause quadrupole sound can usually be neglected. The formulation given by Succi [198] has been used for the calculation of low-frequency noise from wind turbines which is caused by the presence of the tower [16], [84]. An example is given in Section 5.2.

Non-Deterministic Noise, Subsonic Turbulent Flows

Typical examples: High frequency noise from fans or wind turbines.

Mechanism: Primary sources are the fluctuating Reynolds stresses (Lighthill tensor) ⇒ **quadrupole sound**.
In the presence of rigid surfaces, the sound is reflected, scattered, or diffracted at the surface. This leads to ⇒ **additional dipole sound**.

A direct computation of the turbulent flow which causes the noise radiation is not possible. Therefore, a semi-empirical approach is followed. The dependence of the sound intensity on some important parameters such as a typical velocity (e.g. eddy convection velocity) and a typical length scale (e.g. boundary layer thickness) is determined for a simplified but representative case. Examples are given in Section 3.7.2. The empirical input consists of data which have been determined by controlled wind tunnel experiments.

Based on the theoretical scaling laws, the data are analyzed and a universal spectrum shape is deduced which forms the basis of the noise prediction method. An examples for a semi-empirical approach is the model by Brooks, Pope, and Marcolini [30], which is described in more detail in Section 5.3.4.

Deterministic Noise, Transonic Flows

Typical examples: Noise from transonic propellers, high-speed-impulsive noise from helicopters.

Mechanism: The displacement of the air through the motion of the body gives rise to ⇒ **monopole sound**.
Forces act from the body onto the fluid ⇒ **dipole sound**.
The non-linear terms which result from high-Mach number flows with shocks cannot be neglected. These terms give rise to ⇒ **quadrupole sound**.

Simulations of transonic flows require an appropriate finite element or finite volume technique which implies a mesh in order to solve the non-linear Euler or Navier-Stokes equations. The calculation of far-field noise is difficult because the mesh has to include the desired observer position which can lead to numerical inaccuracies and enormous computation times. Therefore, a mixed procedure is used.

The generation of noise, for example by shock waves at the tips (high-speed-impulsive noise), is calculated within the mesh by solving the Euler or Navier-Stokes equations. A control surface (Kirchhoff surface) is defined at a distance which is far enough that all non-linear terms which act as quadrupole sources are included by the surface. The surface is covered with acoustic dipoles and monopoles with respective strength given by the pressure and its normal derivative at the surface. The far-field radiation is determined in the time domain at arbitrary observer locations by applying formulations for rotating Kirchhoff surfaces which have been derived, e.g. by Morino [167] and Farassat [59].

3.7.2 Relevant Cases for Wind Turbine Noise

The flow at wind turbine blades is in the low Mach number range (M = 0.1–0.3) and non-linearities like shocks do not occur. However, as the blade moves through the air it encounters inhomogenities in the incoming flow which are due to the atmospheric turbulence. This inflow turbulence induces an unsteadiness in the flow around the surface that leads to the radiation of sound. Furthermore, the boundary layer on the surface is normally turbulent. This turbulence radiates quadrupole sound which in turn is scattered at the trailing edge. The latter gives rise to an intense noise radiation which is called trailing-edge noise.

This section introduces a few basic cases which are important for wind turbine noise and gives for each case the dependence of the sound intensity on the main parameters such as flow velocity, length scales, turbulence intensity, etc.

Noise from Free Turbulence. The noise radiation originating from free turbulence, for example in a jet, has been a major topic of research since Lighthill's pioneering work [139], [140], [141]. With the concept of quadrupole radiation, which is caused by the presence of fluctuating Reynolds stresses, Lighthill deduced his wellknown eighth power law for the dependence of jet noise on the Mach number of the flow. He found that the acoustic intensity shows the following dependence on the main parameters

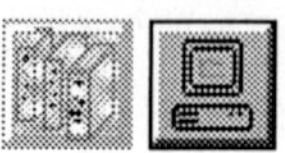

$$I \propto \rho_0 c_0^3 M^8 \left(\frac{l}{r}\right)^2 \alpha^2 . \tag{3.93}$$

Here $M = U/c_0$ is the Mach number with U being a typical velocity, l is a typical dimension in the turbulent region, α is the normalized turbulence intensity, and r the distance to the observer.

Inflow-Turbulence Noise from a Compact Airfoil. Consider a flat plate airfoil of area A_b with mean inflow velocity U (see Figure 3.12). The velocity fluctuates harmonically in the vertical direction with amplitude u and frequency ω

$$u(t) = u_\omega \cdot e^{-i\omega t} . \tag{3.94}$$

Horizontal fluctuations are not considered because they are generally less effective in producing inflow-turbulence noise [155]. If the wavelength of the disturbances $\Lambda = 2\pi U/\omega$ is much larger than the dimensions of the flat plate, the unsteady force can be computed using quasisteady analysis. The unsteady velocity u causes a periodic change in angle of attack. The amplitude of the resulting fluctuating force at the blade section is given by

$$f_\omega = 2\pi \frac{u_\omega}{U} \cdot \frac{\rho_0 U^2 A_b}{2} \cdot \mathcal{S}(\omega) . \tag{3.95}$$

where $\mathcal{S}(\omega)$ is the low-frequency approximation of the incompressible Sears function

$$\mathcal{S}(\omega) = \frac{1}{\left(1 + 2\pi\hat{k}\right)^{0.5}}, \qquad \hat{k} = \frac{\omega C}{2U} . \tag{3.96}$$

Since the wavelength of the radiated sound is larger by a factor $1/M$ than the wavelength of the disturbances, the flat plate can be regarded as compact sound source, i.e. as acoustic point dipole with the dipole strength being equal to the fluctuating force. Application of equation (3.57) yields the sound field

$$\left|p'_\omega(r,\Phi)\right| = \cos(\Phi) k_0 \frac{\rho_0 U^2 A_b}{4r} \cdot \frac{u_\omega}{U} \cdot \mathcal{S}(\omega) \tag{3.97}$$

Considering that a characteristic wave number k_0 is of order $M \cdot 2\pi/\Lambda$, where Λ is a characteristic length scale of the turbulence, the intensity I shows the following dependence on the main parameters

$$I \propto \rho_0 c_0^3 M^6 \left(\frac{A_b}{\Lambda r}\right)^2 \alpha^2 \cdot \cos^2(\Phi)\,. \tag{3.98}$$

Here α denotes again the normalized turbulence intensity. Equation (3.99) shows that the acoustic intensity for a compact airfoil in a turbulent stream varies with the *sixth* power of the Mach number. Recall that the sound intensity for free turbulence varies with the *eighth* power of the Mach number [139]. The directivity factor in equation (3.98) shows that inflow-turbulence noise from compact airfoils has a dipole directivity distribution (see Figure 3.4).

It has to be emphasized that this simple analysis is valid only for very low frequencies. For a typical chord of C = 1 m and a free stream velocity of U = 50 m/s, the frequency should be much less than 50 Hz. Grosveld's approach to model inflow turbulence noise by a single acoustic point dipole located on hub height is therefore not justified [82] (see Section 5.3).

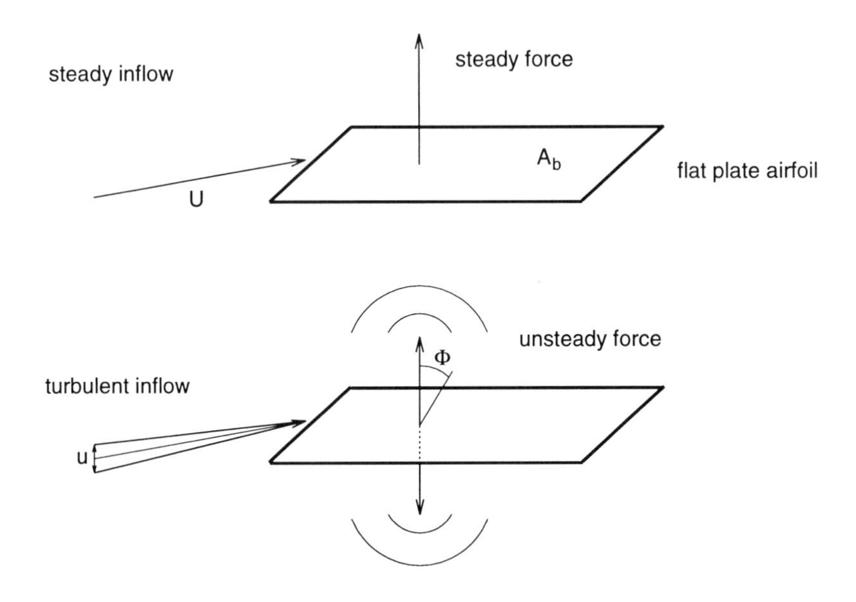

Figure 3.12: Inflow-turbulence noise from a compact airfoil.

The Interaction of Turbulence with Edges. The important noise sources on wind turbines in the high-frequency region are trailing-edge noise and inflow-turbulence noise. Both are caused by turbulence which originates either from the boundary layer around the airfoil or from the atmosphere. As stated in section 3.5, turbulence is an inefficient radiator of sound, particularly at low

  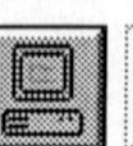 

Mach numbers, meaning that only a relatively small amount of energy is radiated away as sound from the region of the turbulent flow. However, if the turbulence interacts with an edge in the flow – the leading edge for inflow-turbulence noise and the trailing edge for trailing-edge noise –, scattering occurs which changes the inefficient quadrupole radiation into a much more efficient dipole radiation.

For trailing-edge noise, this problem has been treated the first time by Ffowcs Williams and Hall [66]. They investigated the sound field due to a compact turbulent eddy which was modeled by a point quadrupole in the vicinity of a scattering half plane. They found that the far-field intensity of the longitudinal quadrupoles, which are perpendicular to the edge, and the lateral quadrupole, which lies in the plane perpendicular to the edge, are enhanced in intensity by a factor of $(2k_0r_0)^{-3}$. r_0 is the distance from the edge and is supposed to be much less than an acoustic wavelength. They further found that the intensity depends on the *fifth* power of the flow velocity.

The basic problem of turbulence which convects over an edge is treated more generally in Blake [23]. The acoustic analogy is used in the form proposed by Howe [105] with the stagnation enthalpy B as acoustic variable and the source term $\nabla \cdot (\vec{\omega} \times \vec{u})$ (see also Section 3.5.2). A turbulent stream of height l and width s is assumed to convect at mean velocity U making an angle $\overline{\theta}$ with the edge (see Figure 3.13).

The scattering at the edge is computed by using a Helmholtz integral equation and a Green's function for the half plane [162]. The resulting acoustic intensity is found to depend on the following parameters

$$I \propto \rho_0 c_0^3 \cos^3(\overline{\theta}) M^5 \frac{sl}{r^2} \alpha^2 \cdot \sin(\varphi) \sin^2(\theta/2) . \tag{3.99}$$

Here $M = U/c_0$ is again the Mach number with U being a typical velocity in the stream, r is the distance to the observer, and α the normalized turbulence intensity (see Figure 3.13). φ and θ are angles which describe the observer position. An important assumption is that the size of a typical eddy scales on the typical dimension l of the turbulent region. Note that the result found by Ffowcs Williams and Hall is identical to equation (3.99) [66].

Equation (3.99) shows, that trailing-edge noise and high-frequency inflow-turbulence noise have a M^5 dependence on the Mach number. For this reason, noise radiation from edges dominates for low Mach numbers. Equation (3.99) is the basis of several prediction models for trailing-edge noise which are described in Chapter 5 ([30], [82], [156]). In these models the boundary layer thickness δ or the displacement thickness δ^* are taken as a measure for l. Amiet's models for high-frequency inflow-turbulence noise has the same M^5 feature and is outlined in Section 5.3.5.

The theory of trailing-edge noise has been reviewed by Howe [106]. He deduced an expression which gives the sound intensity as a function of the wave vector frequency spectrum of the surface pressure fluctuations (see Section 3.2.2).

The main simplification in the model is the assumption of an infinitely extended plane – i.e. a non-compact chord – with a vanishing thickness which results in a rather simple directivity distribution with maximum radiation in the direction of the plane (see Figure 3.13). Figure 3.14 shows the influence of finite chord and airfoil shape for a turbulent eddy (point quadrupole) in the vicinity of the trailing edge of a NACA 4412 profile ($k_0C = 0.3$). Now maximum radiation occurs above and below the plane of the profile and no sound is radiated in front of the leading edge. The results in Figure 3.14 have been obtained with a two-dimensional implementation of the Helmholtz integral equation (3.89).

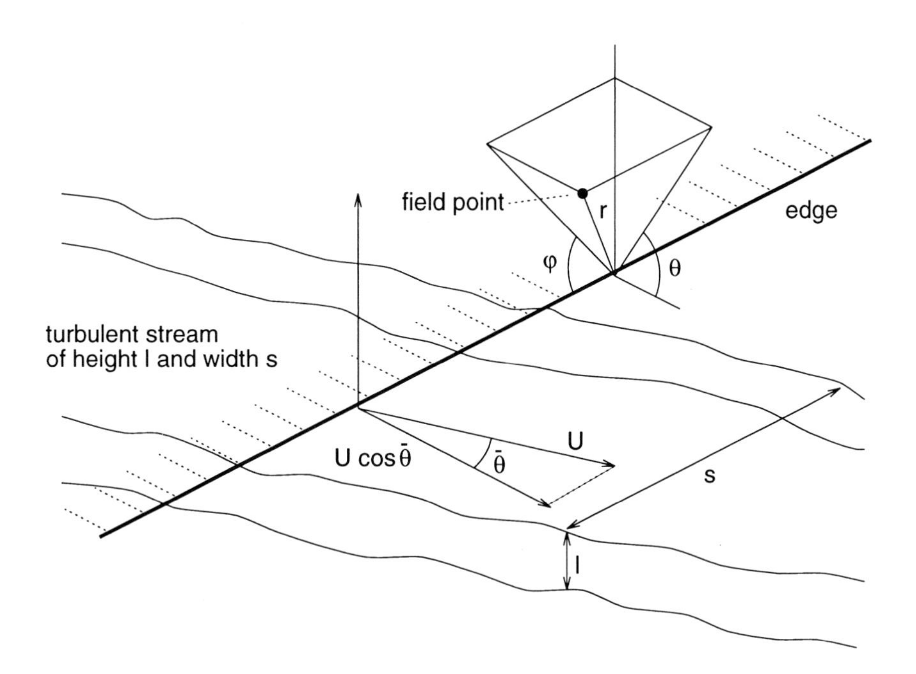

Figure 3.13: Geometry used for the modeling of the interaction of turbulence with an edge [23].

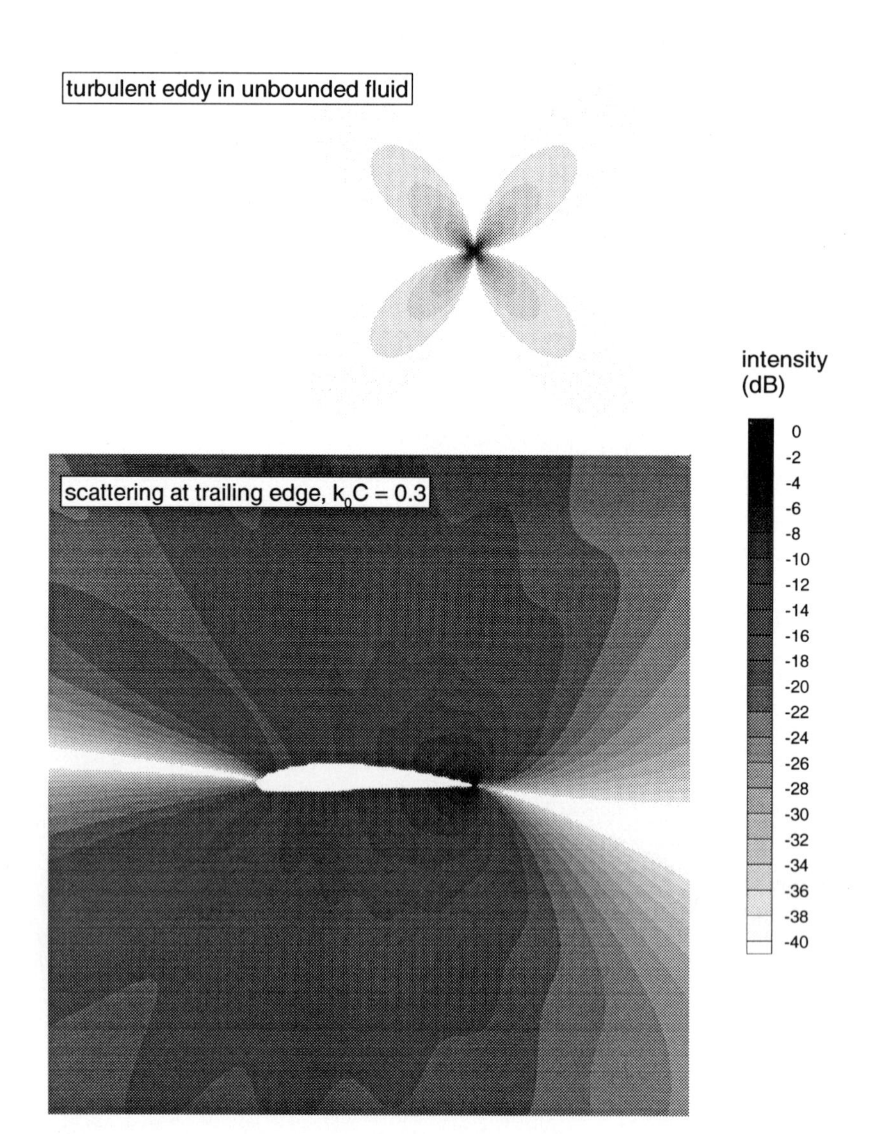

Figure 3.14: Scattering of a quadrupole field at the trailing edge of a NACA 4412 airfoil.

3.8 Conclusions

This chapter has given an introduction to some aspects of aeroacoustics which are important for the understanding, prediction, and reduction of aerodynamic sound.

- Mathematical expressions for the generation and propagation of aerodynamic sound have been derived from the basic conservation laws of fluid dynamics.
- It has been shown that turbulent flow fields are acoustically equivalent to volume distributions of quadrupoles which are due to the fluctuating turbulent Reynolds stresses.
- It has been shown that solid, moving surfaces are acoustically equivalent to surface distributions of monopoles and dipoles with the respective strength given by the local acceleration of the surface and the force acting onto the fluid.
- The difficulties and limitations of noise prediction have been exemplified for a few cases which are important for airfoil noise.

The mechanisms of noise generation which are relevant for wind turbines are explained in the next chapter. An overview of the currently available noise prediction models is given in Chapter 5.

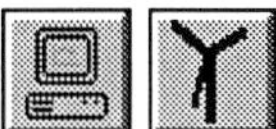

4 Noise Mechanisms of Wind Turbines

This chapter describes the principal noise mechanisms related to wind turbines. After a general discussion of all wind turbine noise sources – including machinery noise and aerodynamic noise – the mechanisms of each aerodynamic noise source are dealt with in more detail. Thus, the spectral properties, their relative significances compared to other noise mechanisms, the key parameters affecting them, and possible means for noise reduction are addressed. The topic of noise reduction is discussed in Chapter 8.

4.1 Classification of Noise Mechanisms

Typically, the acoustic power radiated by a wind turbine is about 10^{-7} of the electrical power [154]. Noise emitted from an operating wind turbine can be divided into

- mechanical noise,
- aerodynamic noise.

Mechanical noise originates from different machinery components, such as the generator and the gearbox. The noise is transmitted along the structure of the turbine and radiated from surfaces, for example, the casing or nacelle raft, the tower, and the rotor blades. Aerodynamic noise is radiated from the blades and is mainly associated with the interaction of turbulence with the blade surface. The turbulence may originate either from the natural atmospheric turbulence present in the incoming flow or from the viscous flow in the boundary layer around the blades.

Machinery noise can be reduced efficiently by well-known engineering methods [177], whereas reduction of aerodynamic noise still represents a problem. At present, manufacturers have been able to reduce the mechanical noise to a level below the aerodynamic noise, now creating the situation that

aerodynamic noise is the dominating noise mechanism. This is aggravated by the increasing size of commercial wind turbines, since mechanical noise does not increase with the dimensions of the turbine as rapidly as aerodynamic noise.

4.1.1 Mechanical Noise

Mechanical noise originates from the relative motion of mechanical components and dynamic response among them, for example, in the gearbox. Pinder [177] reports that due to the tonal character of this type of noise, leading to a penalty of up to 5 dB (see DIN 45681), the recorded annoyance is still caused by mechanical noise, although the energy levels are smaller than those of aerodynamic sources.

This would imply that the distance of the turbine to the closest public buildings has to be approximately doubled. Noise reduction in the relevant frequency ranges is possible and can be expected to be in the order of 8–10 dB due to the removal of the 5 dB penalty, and 3–5 dB, due to the effect of reducing the machinery noise contribution to the measured level [177].

Several elements of the wind turbine contribute to the machinery noise. The main sources are

- gearbox,
- generator,
- cooling fans (including generator),
- auxiliaries: oil coolers, hydraulic power packs for blade pitch.

Furthermore, the hub, the rotor, and the tower may act as loud-speakers transmitting the machinery noise and radiating it.

The transmission path of the noise can be air-borne (a/b) and structure-borne (s/b). *Air-borne* means that the noise is directly propagated from the component surface or interior into the air; *structure-borne* noise is first of all transmitted along other structural components, before it is radiated from another surface. Figure 4.1 shows the type of the transmission path together with the sound power levels determined for individual components by sound pressure measurements at a downwind position of 115 m away from a 2 MW wind turbine, neglecting atmospheric absorption [177]. The main source of machinery noise is the gearbox, which radiates from the nacelle surfaces and the machine raft. Pinder states that the ranking of the various contributions is typical and comparable to other turbines.

The gearbox noise is directly related to transmission errors of the pair of gear meshes, their vibration and loading. The reason for the transmission errors are imperfections in the gear pitch, the form of the meshing teeth and distortions due to tooth loading. Doubling the transmission errors, expressed as the deviation of the ideal geometrical position of the gear from the actual position of the gear under operation, would increase the noise by about 6 dB. The relationship of gear design, operating parameters, and noise generation has not yet been fully established.

The noise produced by the gear teeth is not directly radiated but rather transmitted via the bearings of the gearbox casing. Nacelle insulation and vibration isolation between machinery parts and the enveloping nacelle could result in noise reduction up to 15 dB [177]. This includes efficient damping of noise transmission paths, splitting the nacelle casing and inclusion of flexible couplings. Furthermore, helical gears are quieter than spur teeth by about 1 dB per degree of helix angle.

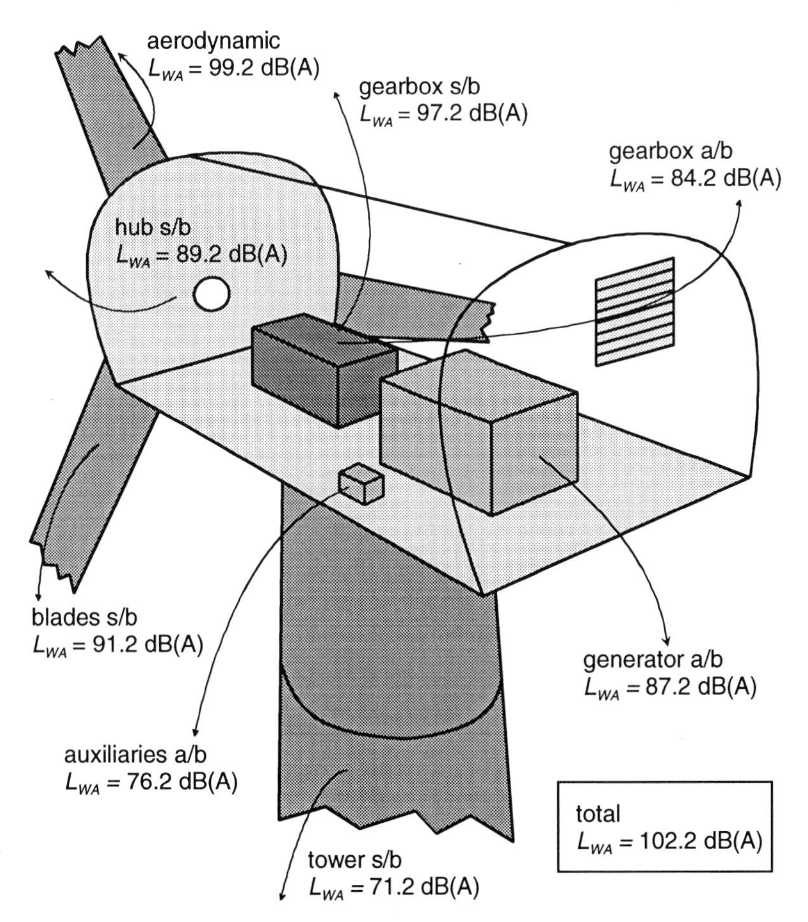

Figure 4.1: Contribution of individual components to the total sound power level of a wind turbine [177].

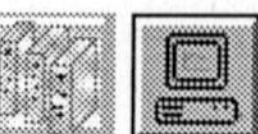

Typical noise spectra for machinery noise are shown in Figure 4.2 and Figure 4.3. The aerodynamic spectrum is typically "smooth", whereas the spectrum of the machinery noise shows a different character: a number of prominent tones, including side bands of meshing or rotation frequency harmonics and components related to the manufacturing cutters, are present. Random distribution of teeth errors will create an additional broadband frequency content [177].

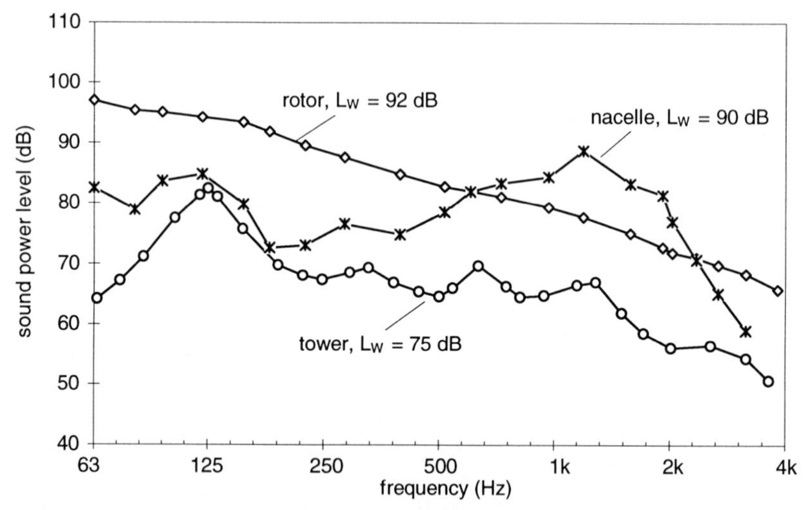

Figure 4.2: Contribution of components to sound pressure level for a 75 kW turbine [177].

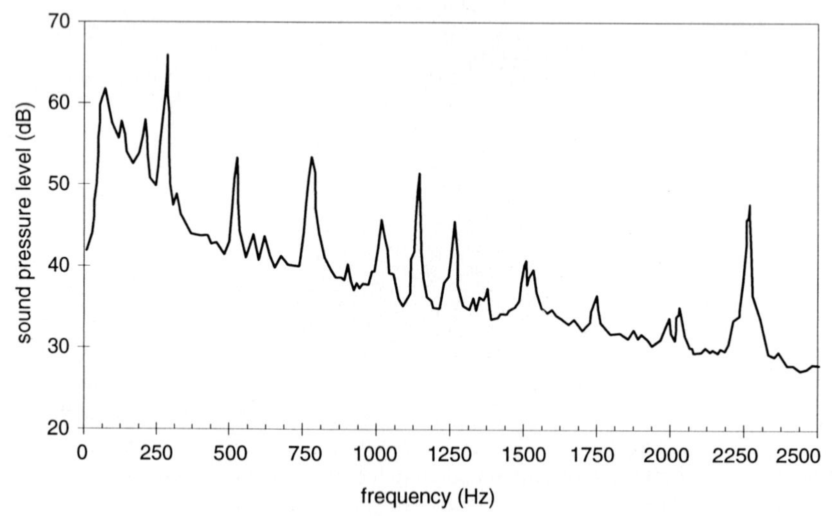

Figure 4.3: Typical sound pressure spectrum including gearbox contribution [177].

4.1.2 Flow-Induced Noise Mechanisms

Within the process of noise generation a large number of complex flow phenomena play an important role. The typical flow environment around a wind turbine blade is shown in Figure 4.4. As for all lifting surfaces, which are operating outdoors, the natural turbulence of the wind is approaching the leading edge of the blade. Initially, the boundary layer, which is developing along the blade, can be laminar; however, for typical Reynolds numbers of approximately $1–5{\cdot}10^6$, transition from laminar to turbulent flow is to be expected.

Close to the leading edge on the upper side of the blade, the flow may strongly be accelerated which can cause a suction peak. Further downstream the flow is decelerating again. This is combined with an increasing adverse pressure gradient which can force the flow to separate from the surface. The flow may reattach further downstream. The deceleration of the flow causes an increase in the boundary layer thickness. Both transition and separation further depend on surface roughness, compliance, and flexibility of the structure. The freestream turbulence and even the ambient pressures can play a role. The boundary layers of the suction side (upper side) and the pressure side (lower side) of the blade together form the wake, which is associated with vorticity shed from the blade.

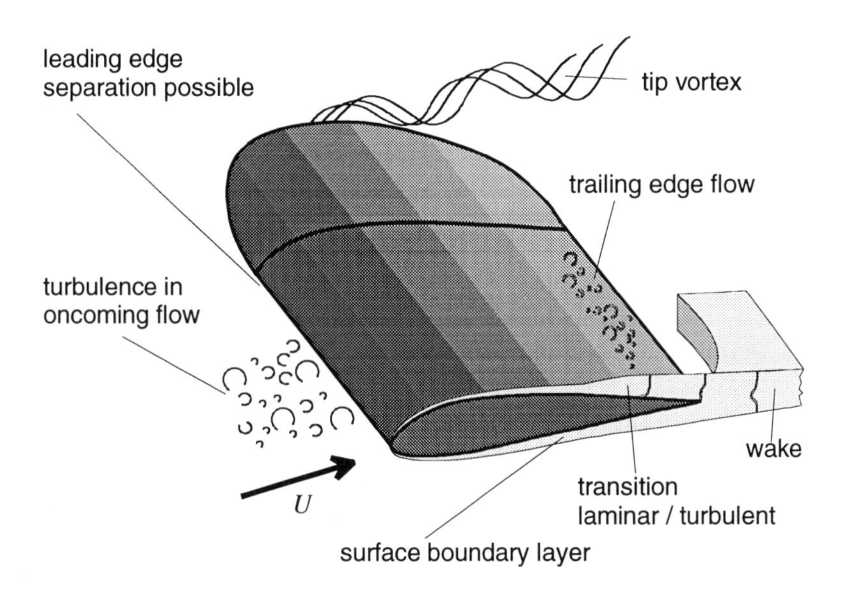

Figure 4.4: Schematic of the flow around the outer part of the rotor blade [23].

For a blade of finite length, the pressure difference between the suction and pressure side tends to compensate at the blade tip. This causes a cross flow over the side edge of the tip which creates a tip vortex that is part of the convected wake.

Table 4.1 shows the various aerodynamic noise mechanisms to be considered, separated into three groups.

Table 4.1: Survey of wind turbine aerodynamic noise mechanisms.

Indication	*Mechanism*	*Main characteristics/importance*
Steady thickness noise / steady loading noise	Rotation of blades / rotation of lifting surfaces	Frequency is related to blade passing frequency (BPF), not important at current rotational speeds
Unsteady loading noise	Passage of blades through tower velocity deficit / wakes	Frequency is BPF-related, small in case of upwind turbines / possibly contributing in case of wind parks
Inflow turbulence noise	Interaction of blades with atmospheric turbulence	Contributing to broadband noise, not yet fully quantified
Airfoil self-noise		
-Trailing-edge noise	Interaction of boundary layer turbulence with blade trailing edge	Broadband, main source of high-frequency noise (750 Hz $< f <$ 2 kHz)
-Tip noise	Interaction of tip turbulence with blade tip surface	Broadband, not yet fully understood
-Stall, separation noise	Interaction of 'excess' turbulence with blade surface	Broadband
-Laminar boundary layer noise	Non-linear boundary layer instabilities interacting with the blade surface	Tonal, can be avoided
-Blunt trailing edge noise	Vortex shedding at blunt trailing edge	Tonal, can be avoided
-Noise from flow over holes, slits, intrusions	Instable shear flows over holes and slits, vortex shedding from intrusions	Tonal, can be avoided

The first group is related to the low-frequency part of the spectrum. This type of noise is generated when the rotating blade encounters localized flow deficiencies due to the flow around a tower, inflow gradients, or wakes shed from the other blades.

Noise due to inflow turbulence constitutes the second group. Depending on the length-scales of the disturbances, the atmospheric turbulence results in global net force fluctuations or local pressure fluctuations around the section which causes relatively low- and high-frequency noise emission, respectively. Up to now the detailed mechanism of noise generation has not been fully understood.

The third group comprises noise generated by the airfoil itself. It is mainly associated with the laminar or turbulent boundary layer on the blade surface. Airfoil self noise is typically of broadband nature, but tonal components may arise due to laminar separation bubbles, blunt trailing edges or flow over slits and holes.

4.2 Low-Frequency Noise

During operation, the rotor blade encounters changes in the flow which are caused by the presence of the tower. Typically, wind turbine towers have a cylindrical cross section that modifies the flow upstream and downstream of the tower. Upstream, the flow is decelerated and can be modeled as a simple two-dimensional potential flow around a cylinder [16], [84]. Downstream, the flow cannot follow the curvature of the tower surface and separates. Depending on the Reynolds number and surface properties, a more or less pronounced wake develops, which involves increased turbulence and a reduction of the mean flow velocity. Thus, the tower causes a reduction of flow speed on both sides, which is, however, much more pronounced downstream of the tower at an equal distance to the tower surface.

As soon as the blade, either located upwind or downwind (Figure 4.5), encounters the flow field generated by the tower, the local angle of attack and the dynamic pressure change and cause a rapid change in blade loading. As seen in Chapter 3.6, forces which act from a surface onto the air are a source of dipole type loading noise. Once the local fluctuations are known a solution of the Ffowcs Williams–Hawkings equation gives the noise. This procedure is described in Section 5.2. Since this mechanism is directly related to the passage of the blades, the spectrum is dominated by the blade passing frequency, f_B, and its harmonics, f_n

$$f_B = n_B \cdot f_R, \quad f_n = n \cdot f_B . \tag{4.1}$$

where f_R is the rotor frequency, and n_B the number of blades. Since these frequencies appear as discrete peaks in the spectrum, this type of noise is often labeled *discrete frequency noise*. For wind turbines, f_B is of the order of 1–3

Hz; therefore, blade-tower interaction contributes only to the low-frequency part of the spectrum. However, the sound pressure levels can be very high, especially for downwind turbines. Typical sound pressure spectra of an upwind and downwind turbine are shown in Figure 4.6. Up to 20 harmonics of the blade passing frequency can be identified in the spectrum of the downwind turbine, whereas the upwind turbine shows no rotational harmonics.

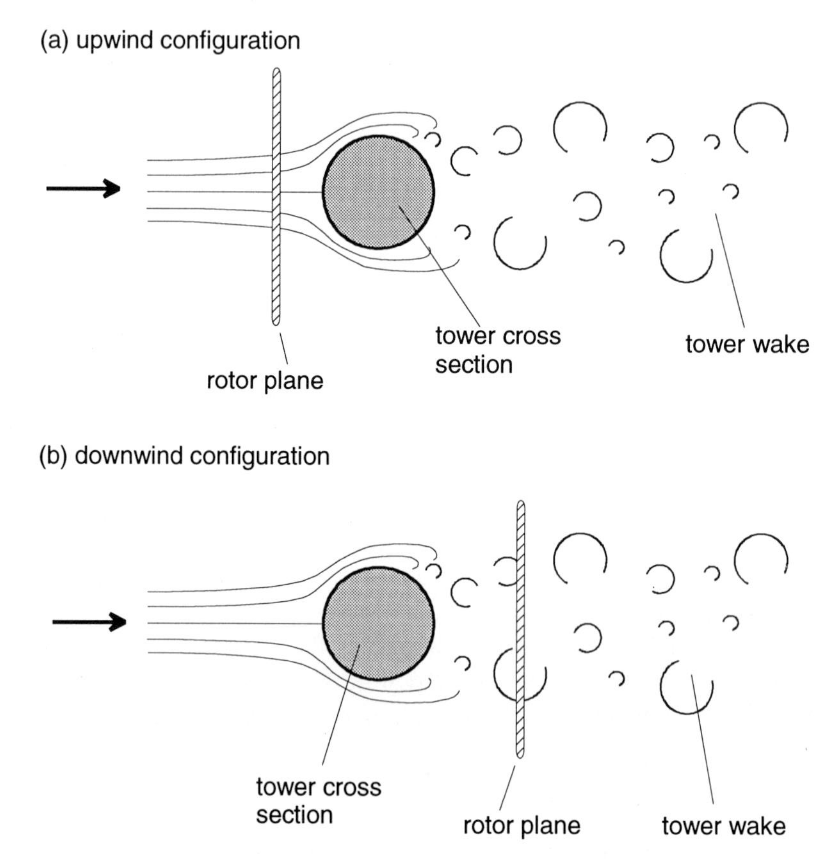

Figure 4.5: Typical flow around a cylindrical wind turbine tower.

Discrete frequency noise is much more important for helicopters, fans, or propellers, with tip Mach numbers > 0.5, where the steady rotating forces can create a strong periodic pressure signature which is called Gutin-noise. Due to the higher rotational speed, the blade passing frequency and its harmonics are shifted up to the audible range of the spectrum. Typical rotor frequencies are given in Table 4.2. Figure 4.7 shows a typical spectrum of a helicopter. Blade vortex interaction (BVI) noise is characteristic for helicopters and originates from the interaction of the rotor blade with its own wake, for example, during descent flight. BVI noise is not important for wind turbines, because of the lesser tip vortex strengths and mainly because the wind transports the wake downstream so that interaction can hardly occur.

Table 4.2: Typical blade passing frequencies (BPF) and Mach numbers of various rotors.

Type of rotor	*Rotational speed* (rpm)	*Blade passing frequency* (Hz)	*Frequency range of discrete frequency noise* (Hz)	*Mach number* (-)
Wind turbine	18–60	1–3	1–150	0.2
Rotorcraft	200–500	10–40	20–200	0.5–1.2
Fan	1000–2000	depends on size, number of blades	depends on size, number of blades	0.1–1.0
Propeller (conventional)	300–1000	30–100	30–2000	0.5–0.8

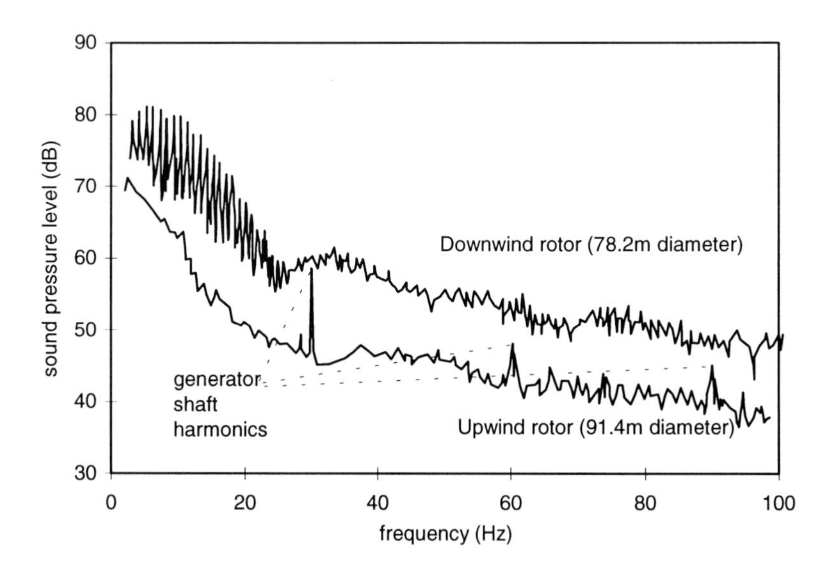

Figure 4.6: Typical low-frequency noise spectrum of an upwind and a downwind turbine [112].

In the case of wind turbines after A-weighting of the spectra, only minor contributions are left from low-frequency noise. Nevertheless, A-weighting considers only the hearing capabilities of human beings, but neglects the other effects that low frequencies may have. The interior organs of humans have low eigen-frequencies in the infrasound region, i.e. below 16 Hz. An excitation of the organs by high sound pressure levels of same frequency may be a cause of annoyance.

Low-frequency noise can also excite vibration of building structures, for example, windows, walls or floors. This can be expected especially when buildings of lightweight construction, such as wooden houses or week-end residencies, are located close to the wind turbine. Figure 4.8 shows sound pressure levels which are sufficient to cause vibrations of building structures. An example of a typical noise spectrum is included for illustration purposes.

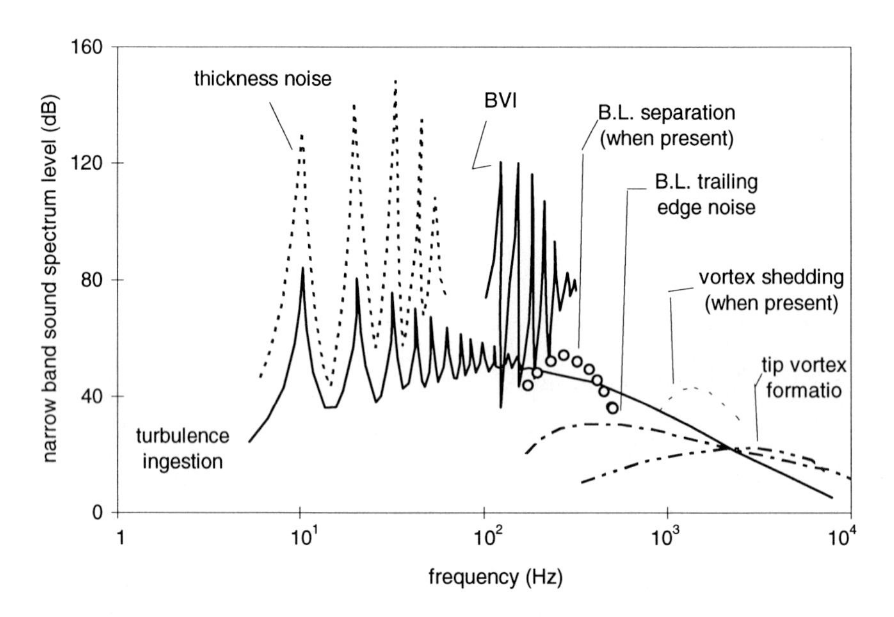

Figure 4.7: Typical noise spectrum of a helicopter [204].

The key parameters which influence low-frequency noise are the orientation and distance of tower and rotor. The higher the distance between tower and rotor, the lower is the disturbance of the flow experienced by the blade and therefore the sound radiation. The noise is reduced considerably if the rotor is placed upwind the tower [84].

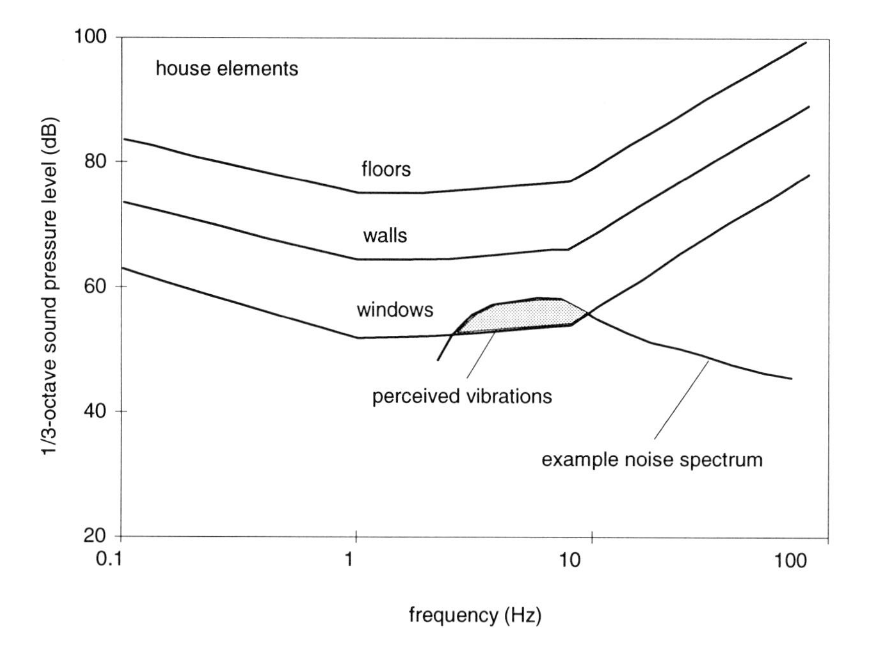

Figure 4.8: Sound pressure level at building resonant frequencies sufficient to cause perceptible vibrations of building structures [196].

4.3 Inflow-Turbulence Noise

Natural atmospheric turbulence encounters the blade and causes a broadband noise radiation. The noise is expected to be very sensitive to the characteristics and properties of the turbulence. Therefore, the properties of atmospheric turbulence will be explained in the following.

4.3.1 Atmospheric Turbulence

Origin of Turbulence. As for every flow over a surface, a boundary layer develops, which in case of the flow over the ground is called the atmospheric boundary layer. Due to the viscosity of the air, the velocity is decreasing continuously with altitude and is zero at the ground. In the boundary layer, the momentum is exchanged by viscous and turbulent shear stresses. The

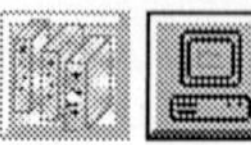

exchange process results in the generation and decay of turbulent eddies which manifest themselves as gusts varying in space and time.

In the so-called *Richardson cascade* the inertial transfer via the turbulence process results in larger scales being successively broken down into smaller scales until the action of viscosity becomes dominant. Sufficiently far away from the ground, the scales must become isotropic, with equal scales in any direction [104].

Driving mechanisms. The cause of the turbulence is twofold: aerodynamic and thermal. Aerodynamic turbulence is generated by the interaction of the flow with the surface, whereas thermal turbulence is generated by the buoyancy of the air due to local heating by the sun. The effects of buoyancy can be neglected at wind speeds above 10 m/s [155]. Each turbulence component is driven by a different mechanism. The longitudinal component, i.e. in direction of the mean wind flow, is mainly driven by the wind shear, whereas the vertical component, normal to the surface, is driven by both, the wind shear and the thermal convection. The scale of the lateral components in space and time can be larger than the longitudinal ones. They can be damped or amplified and undergo interaction with each other.

Eddy sizes and energy spectrum. Although the atmospheric boundary layer contains eddies of all sizes, the dominant ones at height h are found [104] to have scales of order h, i.e. l_{IT} = 100 m (cited in [1]). Figure 4.9 shows the energy spectrum of the horizontal wind speed. A maximum in the spectrum indicates a maximum of the turbulent kinetic energy in the corresponding frequency band.

The first maximum was not measured but estimated and represents the annual seasonal change. The second maximum occurs at 4 days, the typical period for a change in the global weather. The third maximum at 12 hours is associated with the change of day and night and corresponding temperature changes. In the range of 0.1–2 hours an energy gap occurs, because of the absence of relevant processes in the atmosphere. This gap enables wind speed measurements to be performed with a minimum error, selecting averaging times in the order of 10 minutes. The final peak around 1–2 minutes is part of the real gust spectrum. It can be seen that the frequency bands with the most intense energy are located around 0.02 Hz, i.e. 72 1/h, decreasing with rising frequency. The decrease of the power spectral density is described by several models, typically being proportional to $f^{-5/3}$. The typical size for the turbulent eddies is given in Table 4.3 for an assumed chord length of C = 1 m. The values given are based on the paper of Ainslie [1]. A more detailed relationship of eddy size and frequency requires the coherence function of the spatial turbulent velocity components [184].

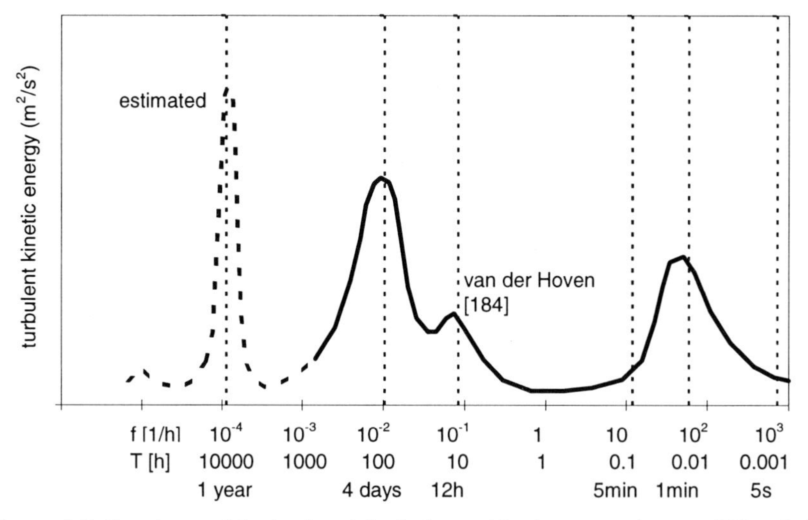

Figure 4.9: Spectrum of the horizontal wind speed in the atmosphere at 100 m height [184].

Table 4.3: Typical size of turbulent eddies.

Typical size Λ of radiating turbulent eddy	$\frac{\Lambda}{\lambda} \approx \frac{U}{c_0} \approx M \approx 0.2 \ldots 0.3,\ M = 0.25$ used				
Typical chord length Speed of sound Wavelength	$C = 1$ m $c_0 = 340$ m/s $\lambda = \frac{c_0}{f}$				
Frequency	(Hz)	10	100	500	2000
Wavelength	(m)	34.00	3.40	0.68	0.17
Eddy size	(m)	8.50	0.85	0.17	0.04
Eddy size / chord length	(%)	850.00	85.00	17.00	4.25

Key parameters. The main parameter describing the turbulence is *turbulence intensity,* i.e. a measure of the turbulence fluctuations. It is defined as the ratio of the standard deviation and the averaged mean velocity of the wind. The key parameters influencing turbulence intensity are the mean wind velocity and the local roughness of the surface. The overall turbulence intensity, which can be measured by a cup anemometer, is dominated by the peak around 0.02 Hz shown in Figure 4.9. However, in case of inflow turbulence noise the

frequency range between 250 Hz and 1000 Hz is dominating. Both frequency ranges are not necessarily correlated.

The stability conditions of the atmosphere are described by the *Richardson Number*, i.e. the ratio of the turbulent energy production due to buoyancy to that due to aerodynamics. The Richardson number is positive for stable or inversion conditions, negative for unstable or convective conditions, and zero for neutral conditions. Lowson [155] cites work performed by Smith and Abbott [161] which demonstrates the influence of the Richardson number on turbulence intensity.

4.3.2 Basic Mechanism

The local flow velocity at the blade is denoted by U. If the length-scale of the disturbance in the atmosphere, i.e. the size of an eddy is Λ, the disturbances occur at a frequency $f = U/\Lambda$, which will be approximately the same frequency $f = c_0/\lambda$ as that of the radiated sound. Depending on the fact whether Λ is larger than the blade chord or smaller, there are two different regimes of inflow turbulence noise. For the following see Section 3.7.2.

Low-frequency inflow-turbulence noise. If the size of an eddy is much larger than the blade chord, the blade will respond with a fluctuation of the total blade loading, see Figure 4.10 (a). This will cause noise radiation of dipole type, varying with M^6. For $U = 50$ m/s and $\Lambda = 10$ m the frequency is $f = 5$ Hz which is in the low-frequency part of the spectrum. The wavelength of the radiated sound is much larger than the blade chord, i.e. $c/\lambda << 1$ [23]. The acoustic wavelength is given by $\lambda = \Lambda / M$, see Table 4.3. Therefore, the blade can be regarded as acoustically compact (see Section 3.5). The latter significantly simplifies the computation of low-frequency inflow-turbulence noise, because the blade can be regarded as an acoustic point dipole with the dipole strength being equal to the net force on the blade.

High-frequency inflow-turbulence noise. If the size of an eddy is comparable or much less than the dimensions of the blade it will induce only local pressure fluctuations which do not affect the global aerodynamic force, see Figure 4.10 (b). Accordingly, the frequency of the radiated noise will be higher and the blade *cannot* be regarded as acoustically compact. The mechanism and the modeling is much more complicated than for the low-frequency part. Turbulent eddies which approach the leading edge are distorted by the mean flow and radiate sound which in turn is scattered at the leading and trailing edge [1]. For details see Blake [23], p.722. This will also result in a different directivity pattern. The airfoil is normally approximated as a (semi-infinite) flat plate. However, considering details of the airfoil geometry, such as thickness, camber etc., will lead to different directivity patterns. In contrast to the low frequencies, the intensity now varies with M^5 instead of M^6 (see Section 3.7.2). The simulation of inflow turbulence noise is significantly complicated if the size of the disturbances is in the same order as the leading edge radius of the airfoil.

Inflow turbulence noise is not yet fully understood and is estimated to be a major source of aerodynamic noise in the frequency up to 1000 Hz. People perceive it as a swishing noise.

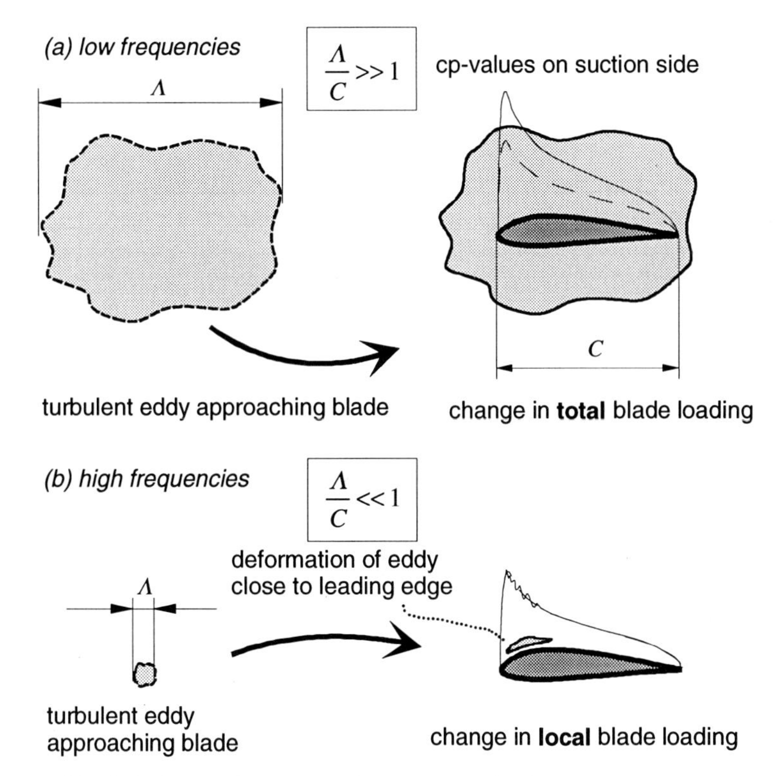

Figure 4.10: Turbulent eddies approaching the rotor blade.

The key parameters influencing the noise generation by inflow turbulence have not yet been fully understood. Lowson states that the stability of the atmosphere and the structure of the turbulence may be an important factor [155].

Within the framework of the DEWI project (see Preface), outdoor measurements on commercial turbines have been performed. The first round of measurements was done on a 500 kW turbine for two wind directions where a varying ground roughness produced different values of turbulence

  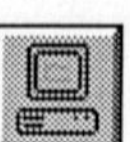 

intensity at 10 m height, namely 18 % for north wind and 12 % for south-west wind. No systematic differences of noise radiation were found in the lower frequency range. However, since it is not known which changes in turbulence intensity occurred at hub height and no information on the spectral content of the turbulence was available, these results are not decisive. A second round of measurements were performed on two smaller machines being located on sites of considerable difference concerning the terrain character, i.e. a turbulence intensity at hub height of 11 % for the first and an estimated value of 25 % for the second site. In contrast to the expected results, the higher turbulence intensity lead to 1–2 dB lower sound pressure levels for frequencies below 1500 Hz. Further measurements are required before drawing firm conclusions [131].

Although different researchers based their models for inflow turbulence noise on a flat plate response to turbulent gusts [3], it is now recognized that the shape of the airfoil and especially the leading edge is of great importance [1], [41]. This offers the possibility to diminish inflow turbulence noise [154]. However, at present it is not yet known definitely which factors influence the noise radiation.

4.4 Airfoil Self-Noise

Even in the case of perfectly steady and turbulence-free inflow an airfoil radiates noise in case instabilities in the boundary layer occur or due to interaction of eddies in the boundary layer with the airfoil surface. Different mechanisms have been identified and are described in the following.

4.4.1 Trailing-Edge Noise

Starting from the stagnation point close to the leading edge, a boundary layer develops on the blade surface. Transition from laminar to turbulent flow occurs at a certain chordwise position, which depends on profile shape, angle of attack, Reynolds number, the structure of the surface, and inflow disturbances. Important parameters describing the boundary-layer turbulence are the length-scale of the energy-bearing turbulent eddies, the turbulent kinetic energy, its spectral decomposition, and the eddy convection velocity. All parameters depend on the normal co-ordinate in the boundary layer. Beneath the boundary layer, the turbulence induces a fluctuating pressure field. Its temporal and spatial transform – the wavevector-frequency spectrum – can be used to compute trailing-edge noise [106].

At low Mach numbers, turbulent eddies, either in free space or on a flat wall, are inefficient sound sources. If there is a sharp edge close to the eddies they will become more efficient as sound sources. Hence, trailing edges of a turbine blade increase the efficiency of the noise radiation from the turbulent eddies in the boundary layer, see Figure 3.14 and Figure 4.11. This is the principal mechanism of turbulent-boundary-layer-trailing-edge-interaction noise which will be called *trailing-edge noise* in the following (see Section 3.7.2).

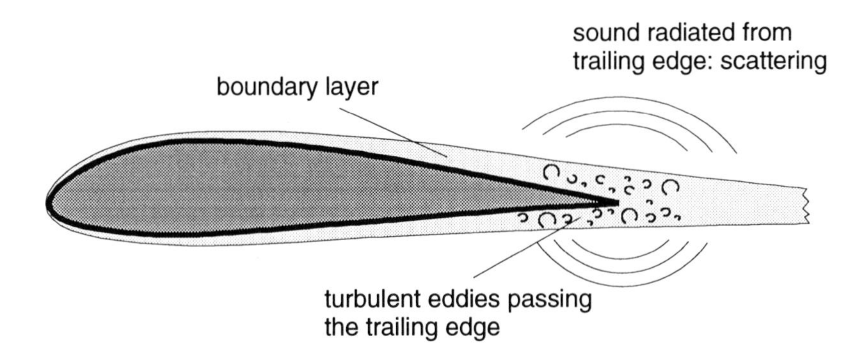

Figure 4.11: Principal mechanism of trailing-edge noise.

Ffowcs Williams and Hall showed that the amplification is strongly influenced by the wave number k_0, i.e. $2\pi/\lambda$, of the emitted sound and by the distance of a turbulent eddy from the trailing edge r_0 (see Section 3.7.2). They found a factor of $(2k_0r_0)^{-3/2}$ by which the acoustic pressure is amplified if the turbulent eddy is close to the edge. They further predict a strong dependency of the noise on the angle between the eddy path and the trailing edge. This implies that noise from a sharp edge surface can be considerably reduced by giving it a swept wing shape [66]. Howe [111] showed that trailing-edge noise can also be reduced by giving the trailing edge a serrated shape (see Section 8.3.2).

People perceive trailing-edge noise as a swishing sound, i.e. broad-band. The peak frequency is typically in the order of 500–1500 Hz, see Section 5.3, depending on the type of turbine and operation. The contribution of trailing-edge noise will dominate in the high-frequency region, if the flow is widely attached over the rotor blade.

The important factors which influence trailing-edge noise are the eddy convection speed and the structure of the boundary layer turbulence close to

the trailing edge, i.e. mainly the distribution of turbulent kinetic energy normal to the surface. The exact shape of the trailing edge is of importance only for relatively high frequencies [109], as well as the properties of the surface, namely the surface impedance.

The reduction of trailing-edge noise is described in detail in Chapter 8. Promising concepts are serrated trailing edges proposed by Howe [111] and confirmed in the wind tunnel by Dassen [41], changes in trailing edge shape (rounding, beveling, etc. [109]), and changes of surface properties (porous trailing edge, variable surface impedance [99]).

4.4.2 Laminar-Boundary-Layer-Vortex-Shedding Noise

If a rotor blade operates at Reynolds numbers less than 10^6, say $10^5 < Re < 10^6$, laminar flow regions on either airfoil side may extend up to the trailing edge. A resonant interaction of the trailing-edge noise with the unstable laminar-turbulent transition can occur. An upstream traveling acoustic wave from the trailing edge couples to the Tollmien-Schlichting instabilities, resulting in tonal noise, see Figure 4.12.

In the case of a laminar boundary layer along the major part of the chord, *boundary-layer instabilities* are likely to occur. These instabilities can lead to separation, a separation bubble and (Tollmien-Schlichting) waves which propagate along the chord. High levels of noise may occur in case the development of instabilities is reinforced (triggered) by the acoustic field and vice versa. This non-linear process (feedback loop) can best be prevented by tripping the boundary layer relatively far upstream of the trailing edge.

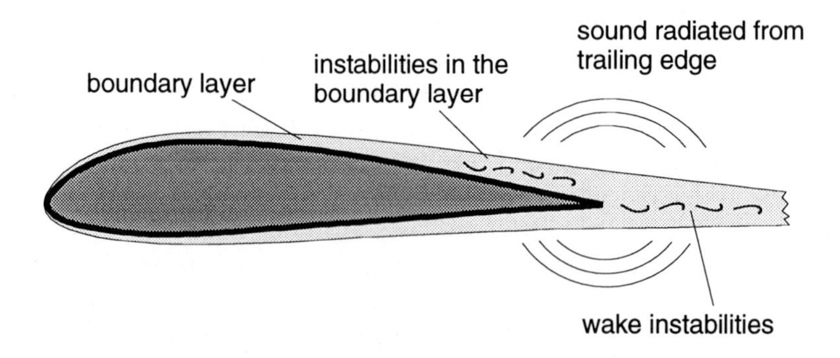

Figure 4.12: Principal mechanism of laminar-boundary-layer-vortex-shedding noise.

Since most modern turbines (500–600 kW) operate at much higher local Reynolds numbers, i.e. $Re > 3 \cdot 10^6$, laminar-boundary-layer-vortex-shedding noise is of minor importance. Only small or medium-sized turbines are expected to produce a significant contribution.

If laminar-boundary-layer-vortex-shedding noise is a problem, a possible solution may be tripping of the boundary layer. The FX 79-W-151A airfoil investigated by Althaus and Würz [2] showed severe tonal noise for angles of attack between 1.5° and 3.8° at $Re < 10^6$. The tones originate from a laminar separation bubble located between 85 % and 95 % chord. After applying a turbulator at 85 %, the tone could be suppressed. Three investigated blade tips (see Chapter 8) showed a similar behavior for comparable flow conditions. Another solution for the elimination of tonal noise is based on a serration at the leading edge which was proposed by Hersh [102].

4.4.3 Tip Noise

Since all noise mechanisms discussed, have a 5th to 6th power dependency on the local flow speed, it is likely that most of the noise is radiated from the tip. However, it is not sure if the tip radiates extra noise due to the three-dimensionality of the flow in this region.

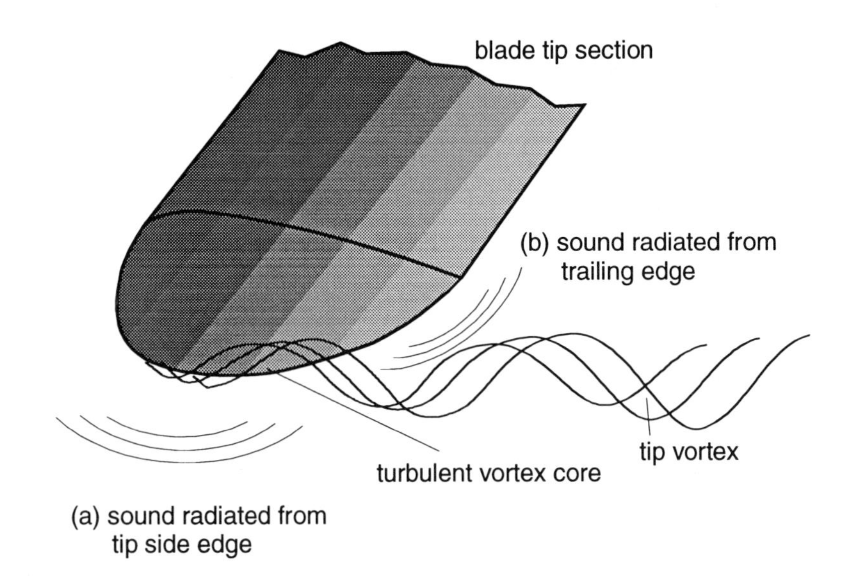

Figure 4.13: Principal mechanism of tip noise.

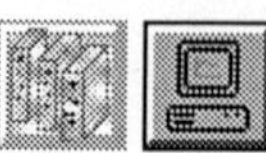

At the tip, the pressure differences between suction and pressure side result in a cross flow over the side edge of the tip which is responsible for the formation of a tip vortex. Brooks and Marcolini [28], [30] suggest that this tip vortex interacts with the trailing edge in the same manner as the boundary-layer turbulence does for trailing-edge noise. Howe [108] states that the noise is created by the cross-flow over the side edge of the tip. Again the mechanism of noise production is similar to the mechanism responsible for trailing-edge noise. Furthermore, separation at the side edge may occur which results in extra noise, see Figure 4.13.

Tip vortex noise is of broad-band character and is assumed to be mainly influenced by the convection speed of the vortex and its spanwise extent [30]. However, it can be expected that the location of the vortex core, the strength of the vortex which depends on the angle of attack, the Reynolds number, and blade load distribution have an influence as well. The importance of tip noise gives rise to controversy. Lowson [155] and Bruggeman [32] summarize some of the different views. The only relevant publication addressing this problem is the one by Brooks, Pope and Marcolini [30]. From their estimation tip noise possibly produces an extra noise of 1–2 dB in some parts of the frequency range.

The proposed mechanisms to reduce tip noise are mainly based on the idea of using a blade tip shape that reduces the interaction of the turbulent vortex core with the edges. Different tip shapes and results from experiments are presented in Chapter 8.

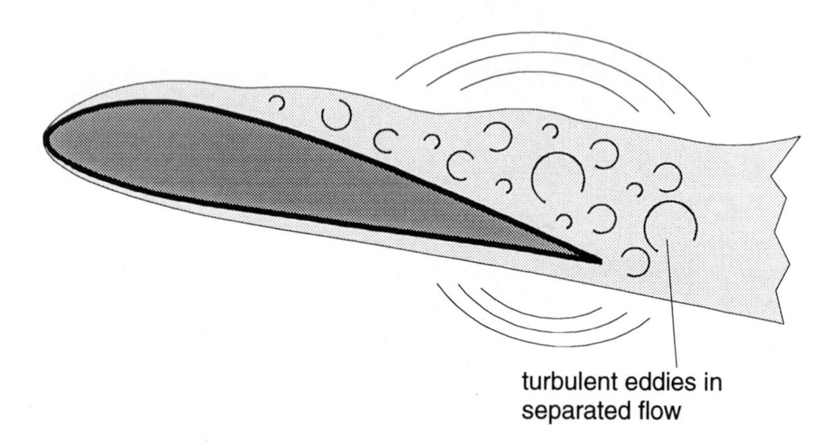

Figure 4.14: Principal mechanism of separated / stalled flow noise.

4.4.4 Separated / Stalled Flow Noise

As the angle of attack increases, stall conditions occur at a certain point causing a substantial level of unsteady flow around the airfoil, see Figure 4.14. Brooks [30] gives an overview of relevant studies. Fink and Bailey [63] found an increase of more than 10 dB for stalled flow relative to trailing-edge noise for low angles of attack. Paterson et al. [173] found that mildly separated flow causes sound radiation from the trailing edge, whereas deep stall causes radiation from the chord as a whole.

Noise caused by stalled flow is of a broad-band nature and is the only major contributing noise mechanism beyond limiting angles of attack [30]. It can only be reduced by avoiding stall conditions at the blade.

4.4.5 Blunt-Trailing-Edge Noise

Blake [23] describes the basic mechanisms of blunt-trailing-edge noise. If the trailing edge thickness t^* of an airfoil is increased beyond a given value, for example, $t^*/\delta^* > 4$ for a NACA 0012 airfoil operating at $Re = 2.63 \cdot 10^6$, a distinct secondary hump becomes apparent in the noise spectrum [23]. Depending on the bluntness and shape of the trailing edge and the Reynolds number, vortex shedding can occur, resulting in a van Karman type vortex street. The alternating vortices in the near wake produce higher surface pressure fluctuations close to the trailing edge (Figure 4.15). As the ratio of t^*/δ^* increases, the band-width of the noise decreases although the continuous character of the spectrum remains basically the same as for a sharp edge trailing edge. Finally, if the ratio of t^*/δ^* is large enough, fluctuating forces will occur resulting in dipole noise of tonal character.

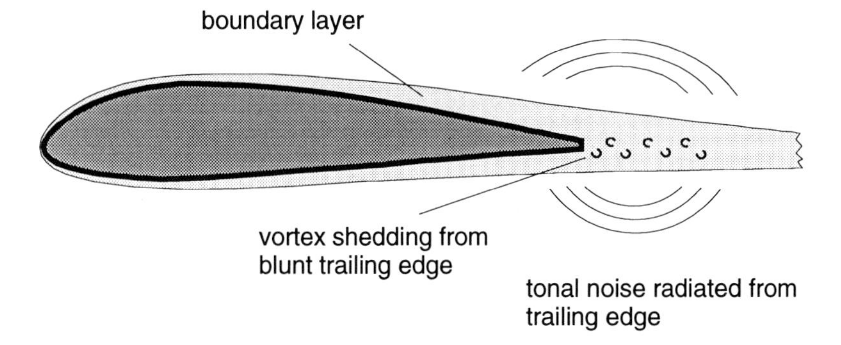

Figure 4.15: Principal mechanism of blunt-trailing-edge noise.

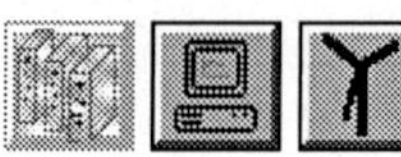

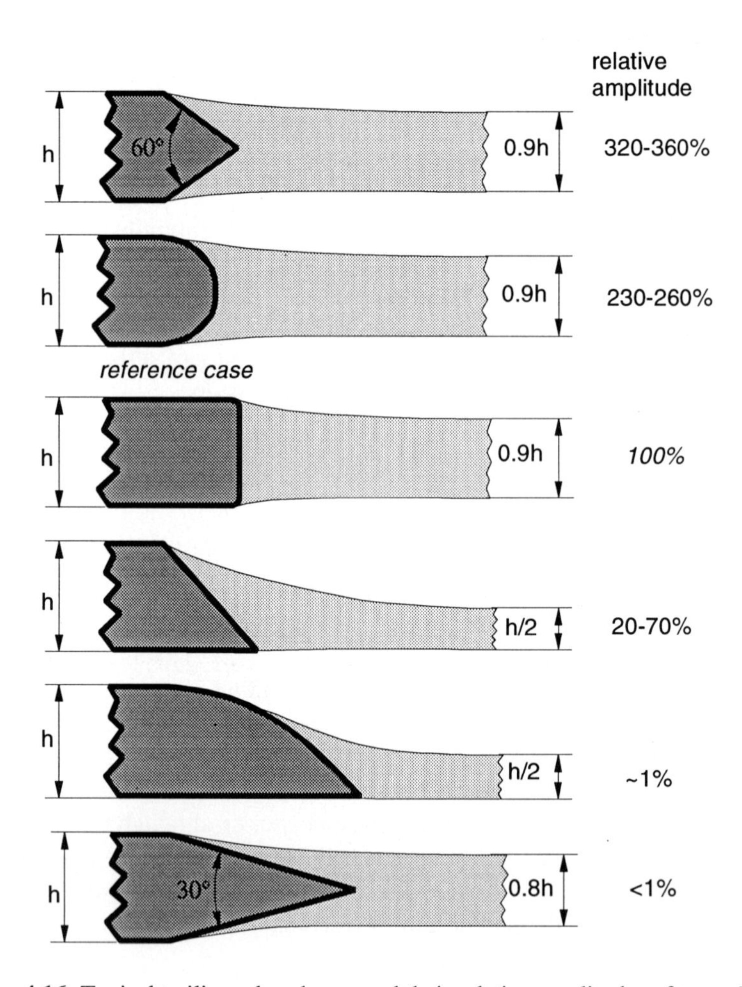

Figure 4.16: Typical trailing edge shapes and their relative amplitudes of tones [23].

As the noise is of tonal character the respective spectrum has a sharp hump shape. The peak frequency depends on the shape of the trailing edge, the Reynolds number, and the so-called *bluntness parameter*, i.e. δ^*/t^*. Blake gives a number of typical Strouhal numbers for different flow regimes and trailing-edge geometries [23]. As an example, Grosveld [82] gives a Strouhal number of $St = f{\cdot}t^*/U_0 \approx 0.1$ for $t^*/\delta^* < 1.3$. From this equation it can be observed that the lower the trailing-edge thickness t^*, the higher the peak shedding frequency f. Sharpening of the trailing edge will therefore shift the peak towards the ultra-sound region. Blake [23] states that in general rigid turbulent-flow airfoils do not generate vortex street sounds when $t^*/\delta^* \leq 0.05$–

0.3. If $t^*/\delta^* \geq 0.3$–0.5, tones are generated and the geometry is the dominating parameter. The practical limit of sharpening the trailing edge is giving by construction, production, and installation considerations, for example, a trailing-edge thickness smaller than 1 mm may not be practical any more, concerning installation, handling, maintenance. Therefore, depending on blade chord and operation conditions a typical trailing-edge thickness is about 1–3 mm.

Figure 4.16 shows typical examples of trailing-edge shapes. They are compared in the form of force or vibration amplitudes relative to those occurring for squared-off blunt edges [23]. It can be seen that depending on the detailed geometry of the trailing edge, an increase of the relative amplitude up to 360 % as well as a decrease down to approximately 1 % compared to the squared-off blunt edge is possible.

4.4.6 Noise due to Blade Surface Imperfections

A manufactured rotor blade operating on a turbine cannot be expected to have the same mathematical and perfect surface as the designed blade shape. Several circumstances can cause deviations during production, assembly, erection, or operation.

The following list gives some examples of so-called blade surface imperfections:

- damage due to
  - installation and erection of the turbine,
  - intense hail,
  - lightning strikes
  - bird impacts,
- dirt due to
  - insects,
  - dust,
  - oil,
- flow imperfections due to
  - loose tapes,
  - slits, for example on partial-pitch turbines,
  - production tolerances.

In general it can be stated that every unwanted disturbance of the flow around the blade may cause additional noise. Typical cases are vortices that generate additional noise while interacting with the blade edges (Figure 4.17). The

noise caused by bluff bodies can be related to their drag coefficient as Nelson and Morfey [169] showed.

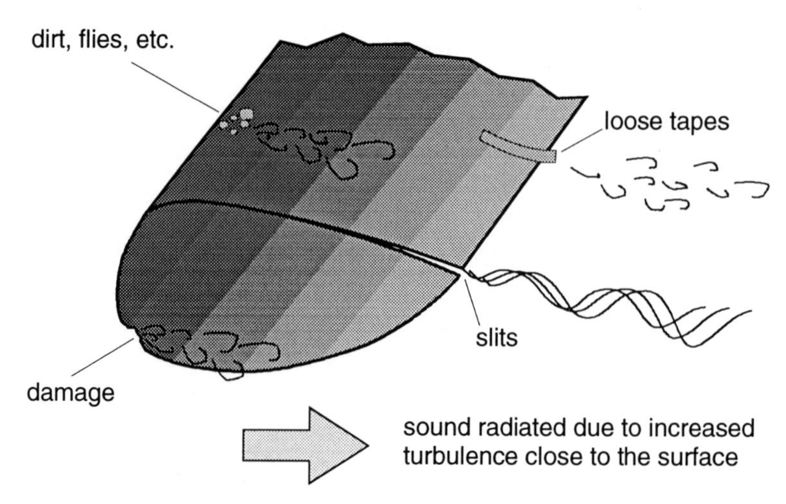

Figure 4.17: Schematic of noise mechanisms due to surface imperfections.

Damage due to lightning strikes can be reduced by installing a lightning protection system. On the other hand damage caused by intense hail and bird impacts cannot be controlled. The same is true for dirt close to the edges resulting from insects in the air. Nevertheless, airfoils which are less susceptible to increased turbulence due to dirt would help to reduce the noise [91].

Slits on the rotor blade are typical for partially pitch controlled turbines. These kind of construction features should be avoided, because it is impossible to guarantee a perfectly closing slit.

4.5 Summary

This section summarizes the different mechanisms of noise generation on wind turbines.

Mechanical noise mostly gives rise to tonal components and can be a dominant noise mechanism under certain circumstances. Mechanical noise will be important in case no means have been applied in order to cut noise transmission paths between machinery components and the structure and / or isolation measures. Typically the main source is gearbox noise. A reduction of noise up to 15 dB can be achieved if appropriate damping is applied.

Low-frequency noise is mainly caused by the interaction of the rotor blade with local flow deficiencies that are caused by the flow around the turbine tower. This phenomenon causes load fluctuations at the blade which are radiated as noise of a dipole character. The noise can be heard as thumping or be perceived as vibrations of building structures. The main influencing parameter is the distance between the rotor relative to the tower and the type of turbine, i.e. upwind or downwind.

Inflow turbulence noise is caused by the interaction of atmospheric turbulence with the rotor blade causing pressure fluctuations at the blade surface. In the modeling of inflow turbulence noise there are two significant limits in which the gust wavelength is either much longer than the size of the body (low frequency) or much shorter (high frequency). These two limits require completely different treatments, which have to be matched at intermediate frequencies. Inflow turbulence noise is regarded as the main noise mechanism for frequencies below 1000 Hz.

Trailing-edge noise is caused by the turbulence in the boundary layer which acts as an acoustic quadrupole radiator. At the trailing edge a strong amplification in sound intensity of this essentially weak acoustic source occurs. Trailing-edge noise is the main aerodynamic noise mechanism for higher frequencies.

At low Reynolds numbers of $10^5 < Re < 10^6$, *laminar-boundary-layer-vortex-shedding noise* can occur. It originates from a non-linear coupling between sound waves of the trailing edge and the laminar-turbulent-transition flow regime. For small and medium-sized turbines, where $Re < 10^6$, this type of noise can become important, whereas it can be neglected for modern turbines.

Tip noise results from the interaction of the turbulent tip vortex with the side edge and the trailing edge of the blade tip. The role of tip noise is not fully understood. Reduction of tip noise is mainly directed towards minimizing the interaction between tip vortex and the blade edges by using appropriate blade planform shapes (see Section 8.4).

Stalled flow noise is caused by the separating flow around an airfoil. Depending on the level of stall, the noise mechanism changes its radiation character from single trailing edge radiation in the case of small-scale separation to full chord radiation in deep stall. It becomes the dominating source beyond a limiting angle of attack. No other means of reduction apart from avoiding stalled conditions is documented in the literature.

Blunt-trailing-edge noise is a result of the vortices shed from the blunt trailing edge. It is of tonal character and can be avoided by sharpening the trailing edge, i.e. typically $t^* \approx 1$–3 mm.

Noise due to surface imperfections on the blade surface has to be split up into two groups, since there is a difference between surface imperfections due to *damaging, ageing, dirt etc.* and *holes or slits* (or even small intrusions) which have been deliberately applied. The latter can be 'well-designed' and the noise can be totally avoided, whereas the former group can only be dealt with by developing blades which are acoustically less-susceptible to these unavoidable imperfections.

Several means of noise reduction of the discussed noise mechanisms are presented in Chapter 8.

5 Noise Prediction

This chapter deals with the prediction of the noise produced by a given turbine under prescribed operating conditions. Beginning with simple rules of thumb for the overall sound pressure level at a given location, models for noise prediction of the whole frequency range are presented. Typical results obtained with the models illustrate the range of application and limitations of state-of-the-art noise prediction. The chapter concludes with developments required and recommendations for improved noise prediction codes.

5.1 Introduction

5.1.1 Role of Noise Prediction

Noise prediction codes are used in the design process of a wind turbine chiefly in order to meet the demand for a minimized noise impact. State-of-the-art noise prediction codes are often based on simple scaling of two-dimensional wind tunnel experiments and neglect the true airfoil shape, leading to limited reliability and accuracy. Many parameters playing a role in aerodynamic and structural design of a turbine are not addressed in the acoustic codes. At present wind tunnel measurements, which are often very expensive, are the only reliable source of information for the designer. Nevertheless, solutions to fill this gap are currently being developed (see next section).

5.1.2 Classification of Codes

There exists a variety of different approaches to determine the noise from a wind turbine. As described in Chapter 4, various types of mechanisms are responsible for the overall noise generated by an operating wind turbine.

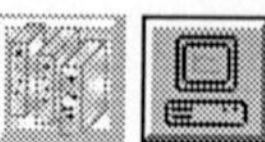

Lowson [155] proposed a classification for the prediction methods (Table 5.1). *Class I* models give simple estimates of the overall sound pressure level as a simple algebraic function of basic wind turbine parameters. These are rules of thumb for easy and fast application. They require only simple input parameters, such as rotor diameter, power, and wind speed. A collection of typical formulas is compiled in Table 5.2 in the next section.

Class II models are most frequently used and represent the state-of-the-art. They are founded on a separate consideration of the various noise mechanisms as depicted in Table 4.1 of Chapter 4. Most of these models are based on the work performed by Grosveld [82], Brooks et al. [30] and Glegg et al. [77], which in turn is based on theoretical analysis [3], [66], [106], see Section 3.7.2. The models contain scaling laws derived from theoretical descriptions and are completed with results of aerodynamic and aeroacoustic measurements. For example, in the case of Brooks et al., acoustic measurements of a NACA0012 profile are related to boundary layer parameters at the trailing edge.

Class III models utilize refined models describing the noise mechanism and relate them to a detailed description of the rotor geometry and aerodynamics. Up to now, no models of this type have been available. Nevertheless, work proposed within the framework of the ongoing JOULE III projects DRAW (JOR3-CT95-0083), STENO (JOR3-CT95-0073), and work undertaken by Lowson [156] are directed towards the development of a *Class III* code.

Table 5.1: Classification of noise prediction codes according to Lowson [155].

Type of code	*Description*
Class I	Predictions giving an estimate of overall level as a simple algebraic function of basic wind turbine parameters.
Class II	Predictions based on separate consideration of the various mechanisms causing wind turbine noise, using selected wind turbine parameters.
Class III	Predictions utilizing complete information about the noise mechanisms related to a detailed description of the rotor geometry, and aerodynamics.

5.1.3 Rules of thumb

This section summarizes currently applied rules of thumb for the prediction of sound power and sound pressure level for a given turbine. The formulas give only a rough estimate of either sound power level or sound pressure level. They are all empirically determined and include relations between noise level

and simple geometrical and/or operational parameters, such as rotor diameter, electrical power output.

Table 5.2: Examples of *Class I* Prediction Formulas.

Reference	*Formula*
[155]	$$L_{WA} = 10\log_{10} P_{WT} + 50 \quad (5.1)$$
cited in [94]	$$L_{WA} = 22\log_{10} D + 72 \quad (5.2)$$
[89]	$$L_{WA} = 50\log_{10} V_{\mathrm{Tip}} + 10\log_{10} D - 4 \quad (5.3)$$
	L_{WA} dB(A) overall A-weighted sound power level V_{Tip} m/s tip speed at rotor blade D m rotor diameter P_{WT} W rated power of the wind turbine
[88]	$$L_{pA} = C_1\log_{10} V_{\mathrm{Tip}} + C_2\log_{10}\left(n_B \frac{A_b}{A_R}\right) + C_3\log_{10} C_T + C_4\log_{10}\left(\frac{D}{r}\right) - C_5\log_{10} D - C_6 \quad (5.4)$$
	If the trailing edge of the rotor blade is sharp
	L_{pA} dB(A) overall A-weighted sound pressure level at a distance of $r_0 = \sqrt{r^2 - h^2}$ C_i dB constants (see below) V_{Tip} m/s tip speed at rotor blade n_B - number of blades A_b m^2 blade area A_R m^2 rotor area C_T - axial force coefficient D m rotor diameter r m distance rotor hub to observer r_0 m distance tower base to observer

Constant (dB)	Hagg [88]	Brochure [176]	De Wolf [45]
C_1	63.3	50	50
C_2	11.5	-	-
C_3	2.5	-	-
C_4	20	20	20
C_5	10	10	10
C_6	27.5	15	12

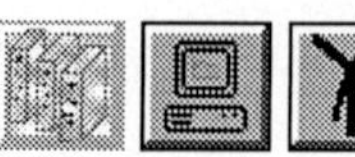

Figure 5.1 shows the sound power levels calculated with different *Class I* models for three WEGA (from the German 'Wind Energie Große Anlagen', i.e. large wind turbines) and one small wind turbine. Only for one turbine, the Richborough 1 MW, the overall sound power level is predicted more or less satisfactorily by the formulas of Lowson [155] and Hau [94]. This example demonstrates the problem of these types of models. Details of the flow around the blade, specific operating conditions, atmospheric conditions etc. are not accounted for in the formulas. Hence, predictions may reflect reality in one case and may deviate significantly in another one.

This is the reason, why *Class I* formulas should only be used to obtain an estimate of the order of magnitude of the overall sound power level or sound pressure level. Results obtained with *Class I* models should be carefully interpreted together with available measurements.

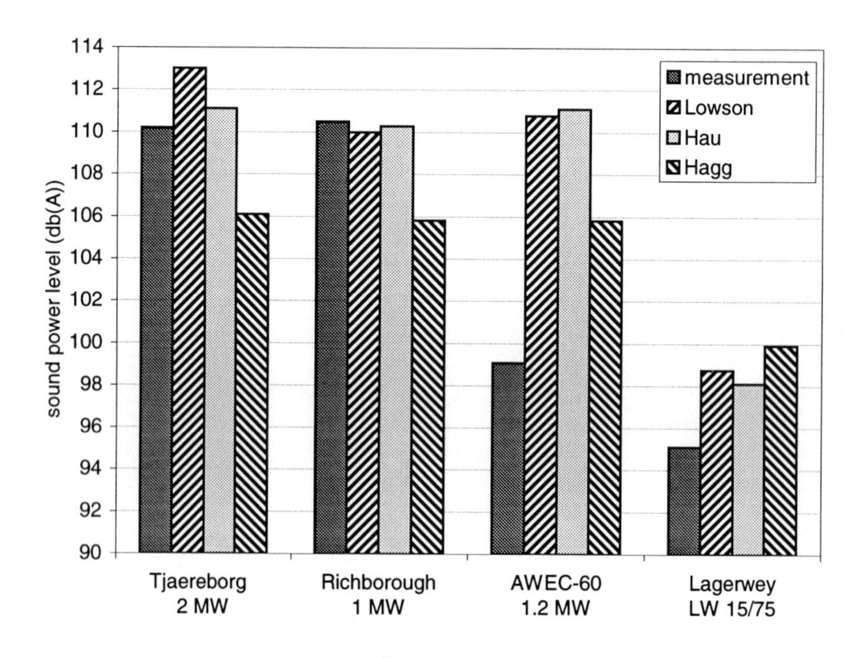

Figure 5.1: Comparison of sound power levels calculated with different *Class I* prediction formulas.

The same is true for the calculation of sound pressure levels. Figure 5.2 shows the sound pressure level as function of observer distance calculated with three different *Class I* models for the Tjæreborg 2 MW turbine. If, for example, a level of 35 dB(A) must not be exceeded at night, the acceptable distances of the closest public buildings will be about 500 m, 600 m or 1000 m depending on the formula applied. These large deviations are not acceptable in practice.

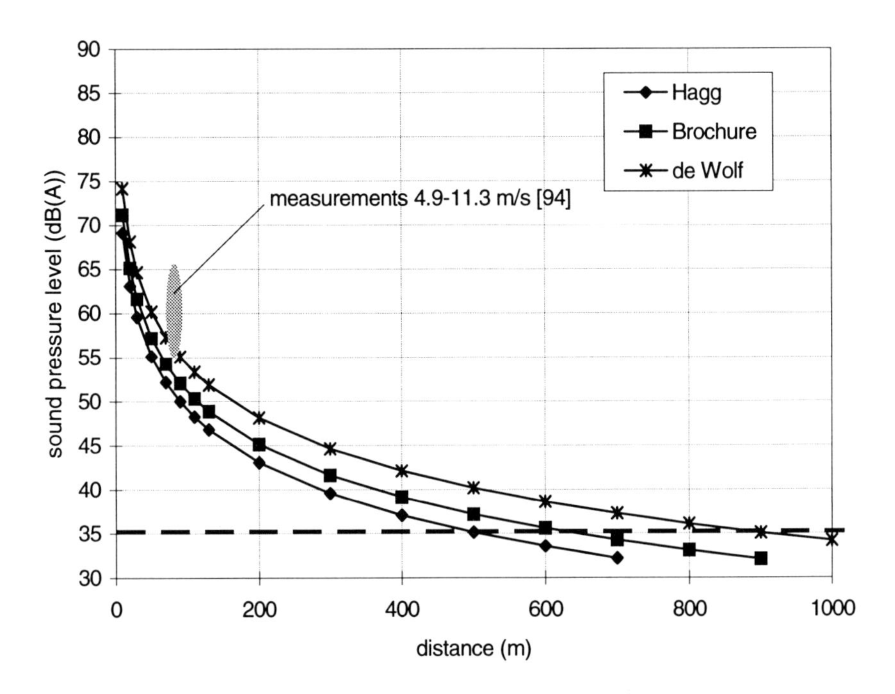

Figure 5.2: Comparison of sound pressure level as function of distance calculated with different *Class I* prediction formulas for the Tjæreborg 2 MW wind turbine; atmospheric absorption is neglected.

5.1.4 Assumptions and Problems of Obtaining Input Data

The development and application of noise prediction codes require a set of assumptions and simplifications to be made. It mainly concerns the geometrical representation of the turbine, the flow around it, its operation, and the dynamic and acoustical behavior of the noise-generating elements. Table 5.3 summarizes typical input parameters for *Class II* models and input parameters possibly used in a *Class III* model to be developed. The main limitations of a *Class II* model are that

- the true airfoil shape for aerodynamic and acoustic calculations is often not considered,
- simple 2D flow is assumed even close to the tips,

– the inflow conditions are often regarded as steady and axial-symmetric.

One of the main problems is that input required especially for *Class III* models, is difficult to obtain. Among the most critical parameters are atmospheric turbulence intensity, surface roughness, geometrical details of the turbine, i.e. tip shapes, tower cross-section and finally wind speed and direction at each blade section. Although, for example, wind speed and direction is often recorded on top of a turbine, the values compared to other parts of the rotor plane may be uncorrelated.

Table 5.3: Typical input for *Class II* and *III* noise prediction codes.

Group	*Parameter*	*Class II*	*Class III*
Turbine configuration	Hub height	x	x
	Type of tower		*x*
	up/downwind orientation		*x*
Blade/rotor	Number of blades	x	x
	Chord distribution	(x)	x
	Thickness of trailing edge	(x)	x
	Radius	x	x
	Profile shape	(x)	x
	Shape of blade tip	(x)	x
	Twist distribution	(x)	x
Atmosphere	Turbulence intensity	x	x
	Ground surface roughness	x	x
	Turbulence intensity spectrum		x
	Atmospheric stability conditions		x
Turbine operation	Rotational speed	x	x
	Wind speed, alternatively: rated power, rated wind speed, cut-in wind speed	x	x
	Wind direction		x

5.2 Low-Frequency Noise

5.2.1 Theoretical Background

Introduction. In Section 4.3.1 the *mechanisms* of *low-frequency noise* (LFN) generation have been discussed. The most important mechanism is blade-tower interaction, i.e. the rotor blade, either positioned upwind or downwind, encounters periodic flow disturbances due to the flow around the tower. The aeroacoustic problem is governed by the Ffowcs Williams–Hawkings equation which has been presented in Section 3.6.1. It describes the generation of sound due to *elementary sources*, namely *monopoles*, *dipoles* and *quadrupoles,* see Sections 3.5 and 3.6. These are related to mass flows (moving volume), forces acting onto the fluid, and turbulent flows (fluctuating Reynolds stresses), respectively.

For the mechanisms of interest and low Mach numbers, mainly monopole and dipole sources are important. The monopole source, the so-called *thickness noise*, represents the effect of the rotor blade volume moving through the air and accelerating with respect to the observer's position. Within this process, the air is displaced for a short period of time and then moves back to its original position. The dipole source, the so-called *loading noise*, comprises the influence of the moving forces acting from the rotor surface on the fluid. The surface pressures are related to the lift produced during the rotor operation in an ambient flow.

The solutions of the Ffowcs Williams–Hawkings equation have been presented in Section 3.6. Since they still contain space derivatives, they are not suited for direct implementation into a computer code. After rewriting the solutions with time derivatives and the additional assumption of *compact acoustic sources*, it is possible to formulate an easily applicable version, see Succi [198] and Farassat [55], [58]. For a typical case of a wind turbine, the assumption of acoustic compactness is valid for frequencies below ~50 Hz. However, the assumption has to be fulfilled by each blade panel. Using the condition of acoustic compactness avoids the necessity of integrating over the whole blade surface. Instead, it is only required to evaluate the formulas at discrete points.

Since aerodynamic codes, e.g. based on the boundary-element method, typically involve a discretization of the blade surface and often assume the local pressure distribution to be constant over the element, the pressure distribution can be replaced by an equivalent force acting at the center of each element. Therefore, the formulation for determining the noise described in the following section is very useful for direct application in a simulation code, because the main input consists only of the forces and geometrical data of source, source motion, and observer.

Compact Noise Source Formulas. The concentrated forces and the blade volume act as compact acoustic sources. Loading noise is produced by two effects, i.e.

- motion of a force with constant magnitude relative to the observer,
- change of the force magnitude and direction with respect to time.

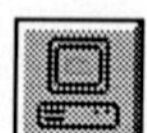

The total acoustic pressure can be obtained by summing up the contributions of all sources n_s, using the formulas given by Succi [198] and Farassat [55], [58]

$$p(\vec{x},t)=\sum_{i=1}^{n_s} p_{\mathrm{t,i}}(t)+p_{\mathrm{ln,i}}(t)+p_{\mathrm{lf,i}}(t)\,. \tag{5.5}$$

The source is located at the position $\vec{y}$ and the observer at the location $\vec{x}$ .The three terms represent the effect of blade thickness $p_t(t)$, blade loading in the near field $p_{\mathrm{ln,i}}(t)$ and the far field $p_{\mathrm{lf,i}}(t)$. The following formulas give the contributions of every acoustic source, i.e. every panel, on the body surface. The thickness term is of monopole character and accounts for the displacement of the air by the blade volume

$$p_{\mathrm{t,i}}(t)=\frac{\rho V_0}{4\pi}\left[\frac{1}{r}\frac{1}{1-M_r}\frac{\partial}{\partial\tau}\left(\frac{1}{1-M_r}\frac{\partial}{\partial\tau}\left(\frac{1}{1-M_r}\right)\right)\right]_{ret}\,. \tag{5.6}$$

The loading term is split into a near-field part

$$p_{\mathrm{ln,i}}(t)=\frac{1}{4\pi}\left[\frac{1}{(1-M_r)^2 r^2}\left(\vec{r}_i\cdot\vec{f}_i\frac{1-\vec{M}_i\cdot\vec{M}_i}{1-M_r}-\vec{f}_i\cdot\vec{M}_i\right)\right]_{ret} \tag{5.7}$$

and a far-field part

$$p_{\mathrm{lf,i}}(t)=\frac{1}{4\pi}\left[\frac{1}{(1-M_r)^2 r}\left(\frac{\vec{r}_i}{c_0}\frac{\partial\vec{f}_i}{\partial\tau}+\frac{\vec{r}_i\vec{f}_i}{1-M_r}\left(\frac{\vec{r}_i}{c}\cdot\frac{\partial\vec{M}_i}{\partial\tau}\right)\right)\right]_{ret}\,. \tag{5.8}$$

It accounts for the noise generated by forces acting on the fluid, e.g. rotation of steady or fluctuating forces located on the rotor blade. The main parameters affecting the noise are the time-dependent force $\vec{f}_i$, the source Mach vector

$$\vec{M}_i=\frac{1}{c_0}\frac{\partial\vec{y}_i}{\partial\tau} \tag{5.9}$$

its first time derivative

$$\frac{\partial\vec{M}_i}{\partial\tau}=\frac{1}{c_0}\frac{\partial^2\vec{y}_i}{\partial\tau^2} \tag{5.10}$$

and the relative Mach number

$$M_r=\frac{\vec{x}-\vec{y}_i}{r}\cdot\frac{1}{c_0}\frac{\partial\vec{y}_i}{\partial\tau}=\vec{r}_i\cdot\vec{M}_i \tag{5.11}$$

which indicates the dimensionless speed of the source in the direction of the observer. Further parameters are the distance to the observer r, air density ρ, speed of sound c_0, and blade segment volume V_0. All terms in square brackets have to be evaluated at retarded time, i.e. the noise is generated at source time τ and reaches the observer at the observer time t. Both times are related by

$$t = \tau + \frac{r(\tau)}{c_0}, \tag{5.12}$$

see also next section and Section 3.4.

Retarded Time Calculation. This procedure can easily be implemented into an unsteady aerodynamic code, because of the explicit time loop arising in such a method. During each time step the relevant geometry of the rotor blades, the wakes, and the loads on the surfaces are known. The procedure can be outlined as follows:

1. Determine the loads on the rotor for each instant of source time τ.
2. Calculate the thickness and loading noise generated at the source time τ using equations (5.6–5.8), resulting in the acoustic pressure signal $p(\tau)$ at source time τ.
3. Determine the current distance of the source to the observer $r(\tau)$ and calculate the observer time t using equation (5.12).
4. After the acoustic pressures are known at each instant of source time τ, sort the acoustic pressures according to their arrival times at the receiver. This gives the acoustic pressure signal $p(t)$ at the receiver time t.
5. Perform a Fourier analysis for one period of the acoustic pressure signal $p(t)$ giving the frequency spectrum of the noise.

5.2.2 Analytical Solutions

Although the formulas of Section 5.2.1 are relatively simple to implement into an existing aerodynamic code, for example, a panel or free-wake method, it is sometimes desirable to study the influence of some fundamental input parameters on the sound pressure levels. The analytical solution described in this section can be used within an engineering code. With relatively simple means, it allows to get an estimate of, for example, the first 10 harmonics of the generated noise.

The first step is to assume that the blades are acoustically compact (see Section 3.5.2) and that the load distribution on a single blade can be represented by a single force. This means the loads are assumed to be concentrated on a reference radius of the rotor blade, typically 72–77 %. The

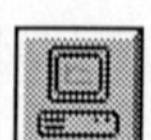

second step is to assume that the loads move in a circle and dynamic movements of the blade can be neglected.

The concentrated forces may then vary during one rotor revolution in an arbitrary way. The time history can be determined with existing aerodynamic codes, for example, by a simple blade-elementum theory (BET) code. Once the time history is available, it has to be expanded into a Fourier series. The Fourier coefficients and basic rotor parameters are the main input for the calculation.

Now an analytical solution can be applied by using equations (5.6–5.8) given by Succi [198] which directly yield the noise spectrum as a function of the load history. The resulting formula is given below. The RMS pressure variation for the nth harmonic of the blade passage frequency is given by the following equation [207]:

$$\hat{p}_n = \frac{K_n\sqrt{2}}{4\pi r}\sum_{m=1}^{\infty}\left[\begin{aligned}&+e^{-im(\phi-\pi/2)}J_{nB-m}(K_nR\sin\gamma)\cdot\left(a_m^F\cos\gamma-\frac{nn_B-m}{K_nR}a_m^M\right)+\\&e^{im(\phi-\pi/2)}J_{nB+m}(K_nR\sin\gamma)\cdot\left(a_{-m}^F\cos\gamma-\frac{nn_B+m}{K_nR}a_{-m}^M\right)\end{aligned}\right]$$
$$+J_{nB}(K_nR\sin\gamma)\cdot\left(a_0^F\cos\gamma-\frac{nn_B}{K_nR}a_0^M\right),\ \text{with}\ K_n=n\frac{n_B\Omega}{c_0} \tag{5.13}$$

The Fourier coefficients of the rotor thrust force

$$a_m^F=\frac{1}{T}\int_0^T F(\tau)\cdot e^{im\frac{2\pi}{T}\tau}\,\mathrm{d}\tau \tag{5.14}$$

and the rotor torque

$$a_m^M=\frac{1}{T}\int_0^T M(\tau)\cdot e^{im\frac{2\pi}{T}\tau}\,\mathrm{d}\tau \tag{5.15}$$

can be determined by a Fourier analysis of the time history of rotor force $F(\tau)$ and rotor torque $M(\tau)$ during one rotor revolution noting that $T = 2\pi/\Omega$. Every m represents one harmonic contributing to the total value of the loads. For m=0 the coefficients represent the mean values of force and torque, i.e.

$$\begin{aligned}a_0^F&=F_{\mathrm{ax},0}\\a_0^M&=M_{\mathrm{torq},0}\end{aligned}\ . \tag{5.16}$$

Finally, the sound pressure level for each harmonic at the prescribed observer position is given by

$$L_{pn} = 10\log_{10}\left(\frac{\hat{p}_n^2}{\hat{p}_{\mathrm{ref}}^2}\right), \text{ with } \hat{p}_{\mathrm{ref}} = 2\cdot 10^{-5} \text{ Pa}. \tag{5.17}$$

The total sound pressure level can be obtained by summing up the pressure of all harmonics, i.e.

$$L_p = 10\log_{10}\left(\frac{1}{\hat{p}_{\mathrm{ref}}^2}\sum_n \hat{p}_n^2\right). \tag{5.18}$$

The elevation angle γ can be determined by $\gamma = \arccos(h/r)$, using hub height h and source receiver distance r, ϕ is the azimuth angle.

Previously, a similar approach has been described by Lowson [148]. The theoretical basis for the noise generated by steady rotating forces was given by Gutin [87] in 1937. Lowson extended the work by Gutin for the noise produced by unsteady forces, see e.g. [78]. The model of Viterna [207] uses the extended Gutin theory and assumes a concentrated rotating force placed at 75 % radius to simulate the low-frequency noise.

5.2.3 Implementation into a Free-Wake Code

The formulas given by Succi in Section 5.2.1 were implemented into the aerodynamic code ROVLM [12], [13], [14], [15]. This code is based on a combined free-wake / hybrid wake method which allows load and flow-field calculations of horizontal axis wind turbines. Figure 5.3 shows a typical example of the wake geometry for a three-bladed turbine generated by the described free-wake code. The problem of high computational effort normally encountered by free-wake methods was solved by freezing the development of the wake in the far field (hybrid wake).

Since the main source of low-frequency is the passage of the rotor blades through the tower wake, an accurate modeling of the flow field around the tower is required for the noise prediction. Depending on the turbine orientation, i.e. upwind or downwind, different models have to be applied (see Chapter 4).

The basic idea of the model is to simulate the presence of the tower by superimposing the flow field of the tower and the flow field around the rotor. Models are incorporated into the code that describe the interaction of the blades with the tower for both, upwind and downwind configuration [16], [83], [84],. Upwind models use the flow around two-dimensional sources or dipoles and downwind models measurements, which can be used for the description of the velocity deficit downstream the tower.

The implementation into a similar aerodynamic code based on the model of vortex particles is described in [208].

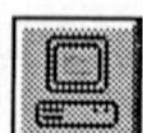

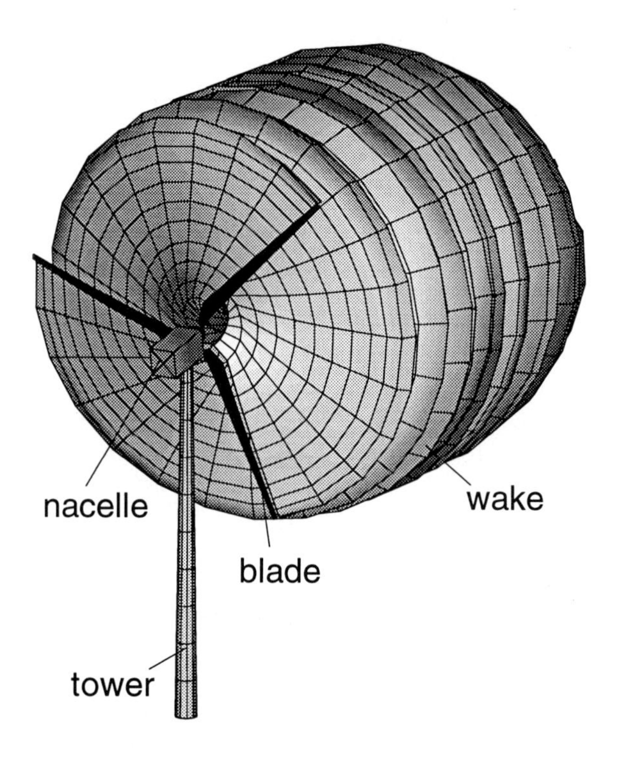

Figure 5.3: Typical example of the wake geometry generated with a free-wake method including dynamic effects.

5.2.4 Typical Results

The methods described above were applied to calculate blade-tower-interaction (BTI) noise of the WTS-4 wind turbine. Comparison with measured data was performed for a 2 MW power output at a wind speed of 12 m/s. The observer was located 91.5 m upwind of the turbine. Calculations were performed with the two models (see above) after generation of the load history (rotor axial force and torque) with the free-wake code ROVLM. The tower wake was modeled by a gaussian shape wake with a wake width of 2 times the tower diameter. The maximum velocity deficit behind the tower was determined according to measurements by Barman on a large wind turbine [17]. Both models use identical load histories as input.

Measured and predicted sound pressure levels show good agreement with the exception of very low frequencies, i.e. below 5 Hz, where ROVLM overestimates the noise (see Figure 5.4).

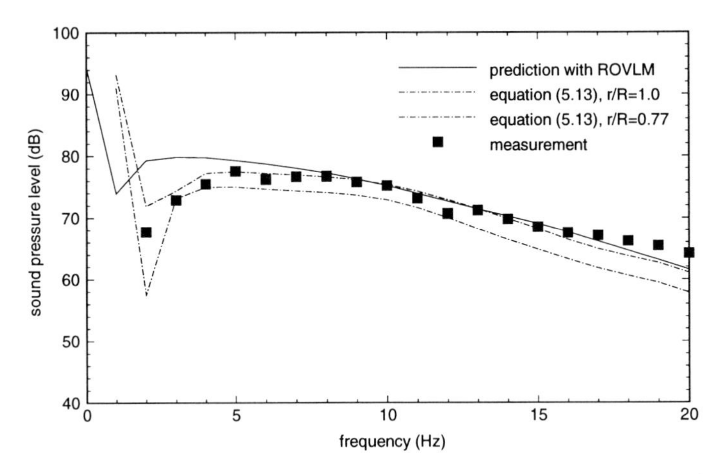

Figure 5.4: Application of equation (5.13) to the case of the WTS-4 wind turbine, $V_W = 12.1$ m/s, $P_{WT} = 2050$ kW, distance = 91.5 m, forces located at 77 % and 100 % radius.

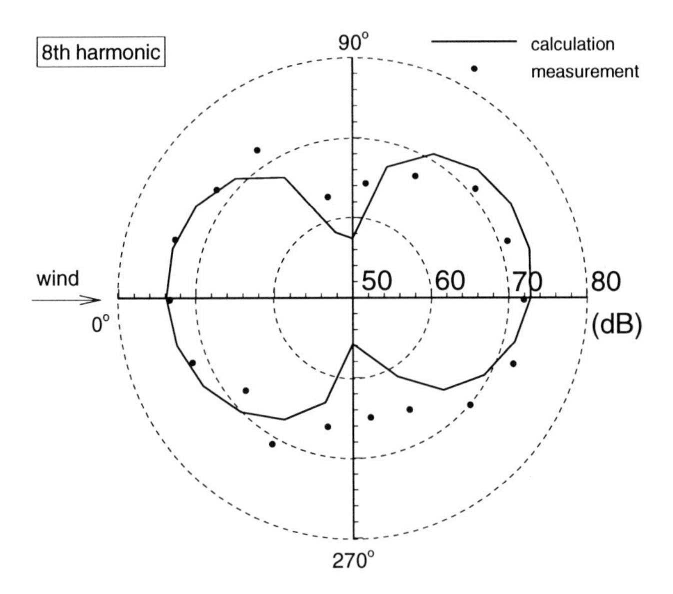

Figure 5.5: Calculated and measured directivity patterns of the WTS-4 wind turbine, $V_w = 12.1$ m/s, $P_{WT} = 2050$ kW, distance = 200 m.

Figure 5.5 shows the measured and predicted directivity pattern of the 8th noise harmonic for the same operation conditions as described above but now at a distance of 200 m. The highest sound pressure levels are found on the axis

of rotation. Agreement in upwind and downwind position is better, i.e. deviation around 2 dB, than in crosswind positions. The measured values exceed the predicted ones by about 8–10 dB in the crosswind position. Shepherd et al. [193], who performed the measurements, found a similar result when comparing their predictions with the measurements. They presented two possible explanations: (a) effects of atmospheric turbulence or (b) blade thickness effects.

5.3 High-Frequency Noise

The prediction models for *high-frequency noise* (HFN) presented here are a selection of the most commonly used codes, being published mainly in the period 1985–1995. Emphasis has been put on models which are directly applicable to the situation of wind turbine noise. If two codes are based on the same model, only the original work is described in detail and its main modifications are indicated for the other one. The models for trailing-edge noise, for example, are quite similar, since nearly all of the codes are based on the work by Ffowcs Williams and Hall [66] or modifications of it (see Section 3.7.2).

5.3.1 Grosveld's Model

Inflow-Turbulence Noise. A well-known inflow-turbulence noise model is provided by Grosveld [82]. The source of this noise is considered to be a point dipole which is located at the hub of the wind turbine. As already discussed in Section 3.7.2 and 4.3.2, this model is only valid for low frequencies, where the lengthscale of the incoming turbulence is large compared to the blade chord.

The structure of the wind and its turbulence are strongly dependent on meteorological conditions, especially on stability conditions. The Grosveld model assumes neutral stability conditions together with a negative temperature gradient as a function of the height above the ground [82]. The total velocity U at a given radial station is determined only from the velocity due to rotation at one reference radius R_{ref}. Wind speed V_w together with induced velocities are neglected. The sound pressure level

$$L_{p,IT} = 10\log_{10}\left(\frac{\overline{w}^2 U^4 CR}{r^2}\frac{n_B \sin^2(\phi)\rho^2}{c_0^2}\right) + K_1(f) + C_1 \qquad (5.19)$$

can then be determined based on the energy spectrum of the acoustic pressures. $K_1(f)$ is a frequency dependent scaling function and C_1 an empirical constant deduced from measurements, see Figure 5.6.

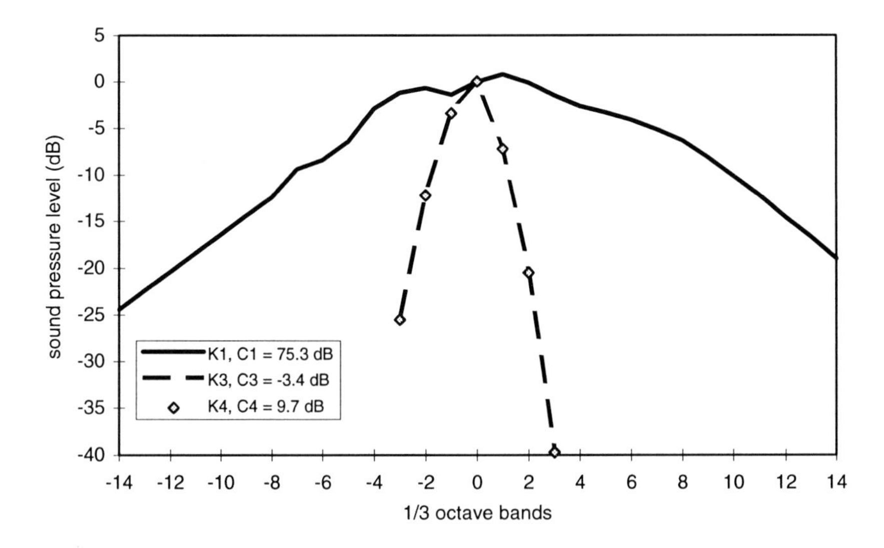

Figure 5.6: Empirical parameters of Grosveld's model [82].

The peak frequency of $K_1(f)$ can be found from

$$f_{peak} = \frac{St \cdot U}{h - 0.7R}, \quad U = \Omega R_{\text{ref}} \tag{5.20}$$

and the Strouhal number $St = 16.6$. The reference chord length of the blade

$$C_{ref} = C_t + 0.3(C_r - C_t) \tag{5.21}$$

is calculated from chord length at tip and root assuming a tapered blade. This corresponds to a location at 30 % of the blade length. The reference turbulence intensity is given by

$$w_r^2 = 0.2\left(2.18 V_w h^{-0.353}\right)^{\frac{1}{1.185 - 0.193 \log_{10} h}} \tag{5.22}$$

and the root mean square turbulence intensity by

$$\overline{w}_r^2 = w_r^2 \left(\frac{h w_r}{V_w R \left(w_r - 0.014 w_r^2 \right)} \right)^{-\frac{2}{3}} . \tag{5.23}$$

The angles ϕ, θ, ψ represent the relative position of the observer to the noise source and therefore the source directivity. For a detailed definition of the angles see [82].

Trailing-Edge Noise. Grosveld's model [82] uses the results of Schlinker and Amiet [186] deduced for helicopter noise. They treated the problem of a frozen turbulence pattern convecting downstream over the trailing edge of an airfoil. The airfoil is modeled as a semi-infinite plate. The pressure jump across the airfoil is forced to fulfill the Kutta condition at the trailing edge, resulting in an induced pressure field which will be radiated from the trailing edge region. The noise is described as a function of local Mach number

$$M = \frac{\Omega r}{c_0}, \tag{5.24}$$

the eddy convection Mach number $M_c = 0.8 \cdot M$, boundary-layer thickness δ, length of blade segment Δs, and observer distance r. The sound pressure level is given by (see equation (3.99) in Section 3.7.2)

$$L_{p,\mathrm{TBLTE}} = 10 \log_{10} \left(\frac{\delta \cdot \Delta s U^5 \overline{D}_1}{r^2} n_B \right) + K_2(f) + C_2 \tag{5.25}$$

containing the frequency dependent scaling function

$$K_2(f) = 10 \log_{10} \left\{ \left(\frac{St'}{St_{\max}} \right)^4 \cdot \left[\left(\frac{St'}{St_{\max}} \right)^{1.5} + 0.5 \right]^{-4} \right\}, \tag{5.26}$$

the Strouhal numbers

$$St' = \frac{f \delta}{U}, \quad St_{\max} = 0.1, \tag{5.27}$$

and the empirical constant

$$C_2 = 10 \log_{10}(3.5) \cong 5.44 \text{ dB}. \tag{5.28}$$

The directivity factors are given by

$$\overline{D}_1(\theta, \psi) = \sin^2(\psi) \cdot \overline{D}_2\left(\theta, \frac{\pi}{2}\right). \tag{5.29}$$

and

$$\overline{D}_2\left(\theta,\frac{\pi}{2}\right)=\frac{\sin^2(\theta/2)}{(1+M\cos\theta)\left[1+(M-M_c)\cos\theta\right]^2} \tag{5.30}$$

Blunt-Trailing-Edge Noise. The model for blunt-trailing-edge noise contains two different flow conditions which are characterized by the ratio of the length over the body from which the vortices are shed, i.e. the trailing edge thickness t^*, and the displacement thickness of the boundary layer δ^* at the trailing edge. If this ratio becomes larger than 1.3, the overall sound pressure level at a given observer position will be given by

$$L_{p,\mathrm{BTE}}=10\log_{10}\left(\frac{U^6 t^* \Delta s}{r^2}\frac{\sin^2(\theta)\sin^2\psi}{(1+M\cos\theta)^6}n_B\right)+K_3(f)+C_3 \tag{5.31}$$

$$(t^*/\delta^*>1.3)$$

and the peak frequency is given by

$$f_{\mathrm{peak}}=\frac{0.25U}{t^*+\frac{\delta}{4}}. \tag{5.32}$$

However, if the ratio is less than 1.3 the sound pressure level will be given by

$$L_{p,\mathrm{BTE}}=10\log_{10}\left[\frac{U^{5.3}t^*\Delta s}{r^2}\frac{\sin^2\left(\frac{\theta}{2}\right)\sin^2\psi}{(1+M\cos\theta)^3\left[1+(M-M_c)\cos\theta\right]^2}n_B\right]+K_4(f)+C_4$$

$$(t^*/\delta^*<1.3) \tag{5.33}$$

which contains two empirically determined functions $K_3(f)$, $K_4(f)$ given in Figure 5.6. The peak frequency is given by

$$f_{\mathrm{peak}}=0.1\frac{U}{t^*}. \tag{5.34}$$

5.3.2 NLR Model

The model of NLR was developed by de Wolf [44] in 1986 based on the work previously published by Grosveld [82]. The respective report is written in

Dutch and will therefore be treated only shortly. The main differences to the Grosveld's model are:

- the local flow conditions at each blade section are determined by a blade-element-momentum (BET) theory code with linearized tip-correction instead of using only the velocity caused by rotation,
- boundary-layer thickness at the trailing edge is determined by a boundary-layer code instead of using the boundary-layer thickness of a flat plate,
- inflow turbulence noise: K_1 is fitted to measurement data of a large series of Dutch wind turbines instead of using the measured data of Grosveld,
- an additional noise source, i.e. inflow turbulence noise caused by blade thickness, is implemented into the code. It accounts for the noise of an unsteady drag force component [76],
- a further noise source, i.e. tip noise, based on the model of Brooks et al. [28] has been implemented,
- for the calculation of inflow turbulence the ESDU data on atmospheric turbulence as a function of surface roughness [52], [53], [54] are used instead of the data given by Grosveld,
- the model allows calculations for various off-axis observer positions, whereas the Grosveld model is limited to positions in vertical planes including the rotor axis.

In 1993, Dassen [40] published a validation of the newest code version RHOAK3 on three different upwind turbines (200 kW, 350 kW, 1000 kW). Based on the validation process, the code was adjusted to better reflect the features of the measurements. The main adjustment comprises that the calculation of the Strouhal number for trailing-edge noise is now based on the displacement thickness of the boundary-layer using a value of $St = f\delta^*/U = 0.1$. In the previous version it was founded on the boundary-layer thickness and using of $St = f\delta/U = 0.4$. Furthermore, the K_1, K_2, K_3, K_4 were fitted to the measured data.

5.3.3 Glegg's Model

The model of Glegg [77] incorporates mainly the following noise sources

- unsteady lift noise,
- unsteady thickness noise,
- trailing-edge noise,
- noise from separated flows.

Following the approach by Amiet [3], Glegg relates the induced aerodynamic pressures on an airfoil to a field of oncoming harmonic gusts, which represents the atmospheric turbulence, that is assumed to be incompressible and isotropic. Aerodynamic and acoustic pressures are then related by using the equation of Ffowcs Williams and Hawkings as given in Section 3.6.1 for the acoustic pressure radiated from a dipole source. This procedure provides a generalized model that is valid for unsteady lift and thickness noise. The corresponding airfoil response functions are evaluated using the work of [3]. In contrast to common Class II models, Glegg considers the finite chord of the airfoil. Noise from bluff bodies, fairings or housings is determined by using the results by Nelson and Morfey [169], who performed acoustic measurements to relate the steady drag forces of a bluff body to the acoustic pressure. The trailing-edge noise model is based on results by Amiet who specifies the acoustic pressure as a function of the surface pressure spectrum of the blade boundary layer [4], [5].

Furthermore, Glegg considers the effect of acoustic scattering that occurs when the blade, acting as a dipole, is passing the tower. Finally, correction formulas were deduced for the results obtained for translational blade motion to account for the effect of blade rotation and the periodic motion of the blade as it passes through turbulence, which is convected with the mean windspeed.

5.3.4 Brooks, Pope, Marcolini Model

Brooks et al. [30] performed a number of aerodynamic and acoustic measurements for a set of seven 2D NACA 0012 profile sections (chord length 2.54 cm, 5.08 cm, 10.16 cm, 15.24 cm, 22.86 cm, 30.48 cm, span 45.72 cm) and a 3D blade tip (chord length 30.48 cm, span 45.72 cm) at Reynolds numbers between 400.000 and 1.500.000 which is quite low for wind turbine applications. Boundary layer thickness, momentum thickness, displacement thickness, and sound pressure levels were determined for tripped (at 20 % chord) and untripped conditions at free-stream velocities up to 71.3 m/s and angles of attack in the range 0°–25°. The models were tested in the low-turbulence potential core of a free jet located in an anechoic chamber. Eight ½" diameter free-field-response microphones were mounted in the plane perpendicular to the 2D model midspan [30]. The extensive measurement program resulted in a set of spectral scaling formulas for the sound pressure levels in the vicinity of a profile section which were implemented into a FORTRAN 77 computer code. The listing of the code is included in the report. The application of the code to a wind turbine requires noise contributions from all the blade segments to be considered.

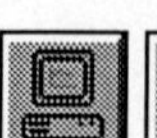

Trailing-Edge Noise. The scaling law applied is based on the result of Ffowcs Williams and Hall [66] for the problem of turbulence convecting past the trailing edge on a half-plane (see Section 3.7.2). The noise is described as a function of local Mach number M, displacement thickness δ^*, length of the blade segment s, angle of attack α and the distance of the source to the observer position r. The total sound pressure level is given by

$$L_{p,\mathrm{TBLTE}} = 10\log_{10}\left\{10^{L_{p,\alpha}/10} + 10^{L_{p,s}/10} + 10^{L_{p,p}/10}\right\} \tag{5.35}$$

with

$$L_{p,\alpha} = 10\log_{10}\left(\frac{\delta_s^* M^5 \Delta s \overline{D}_h}{r^2}\right) + G_B\left(\frac{St_s}{St_2}\right) + K_5 \tag{5.36}$$

representing the effect of angle of attack,

$$L_{p,s} = 10\log_{10}\left(\frac{\delta_s^* M^5 \Delta s \overline{D}_h}{r^2}\right) + G_A\left(\frac{St_s}{St_1}\right) + (K_6 - 3) \tag{5.37}$$

the contribution of the suction side and

$$L_{p,p} = 10\log_{10}\left(\frac{\delta_p^* M^5 \Delta s \overline{D}_h}{r^2}\right) + G_A\left(\frac{St_p}{St_1}\right) + (K_6 - 3) + \Delta K_6 \tag{5.38}$$

the contribution of the pressure side of the airfoil. The Strouhal numbers are defined as

$$St_s = \frac{f\delta_s^*}{U} \text{ and } St_p = \frac{f\delta_p^*}{U}. \tag{5.39}$$

The directivity of the high-frequency noise is given by

$$\overline{D}_h = \frac{2\sin^2\left(\frac{\theta}{2}\right)\sin^2\psi}{(1 + M\cos\theta)\left[1 + (M - M_c\cos\theta)\right]^2}. \tag{5.40}$$

The peak Strouhal numbers are given by

$$St_1 = 0.02 \cdot M^{-0.6}, \tag{5.41}$$

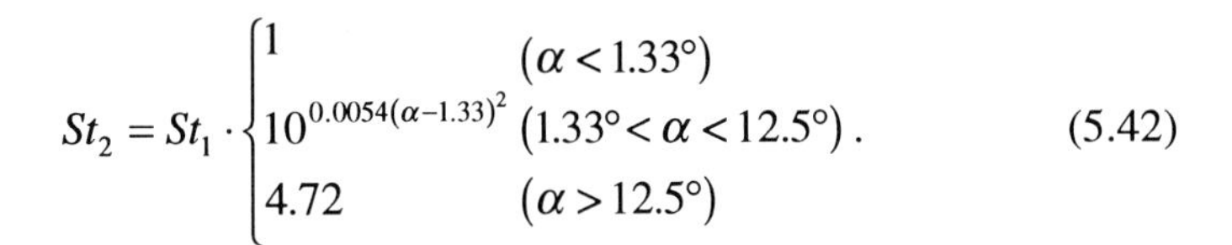

$$St_2 = St_1 \cdot \begin{cases} 1 & (\alpha < 1.33°) \\ 10^{0.0054(\alpha-1.33)^2} & (1.33° < \alpha < 12.5°) \\ 4.72 & (\alpha > 12.5°) \end{cases}. \tag{5.42}$$

Separated / Stalled Flow Noise. The measurements of Brooks et al. [30] were performed for angles of attack from 0° up to 25°. They state that to their knowledge no predictive methods for noise caused by stalled flow have been developed. Hence, the same scaling theory as for turbulent-boundary layer trailing edge interaction noise with some slight modifications has been applied. See the previous section for the respective formulas.

Laminar-Boundary-Layer-Vortex-Shedding Noise. The model for noise caused by laminar boundary layer vortex shedding is assumed to scale similar to noise caused by trailing-edge noise, i.e.

$$L_{p,\mathrm{LBLVS}} = 10\log_{10}\left(\frac{\delta_p M^5 \Delta s \overline{D}_h}{r^2}\right) + G_1\left(\frac{St'}{St'_{\mathrm{peak}}}\right) + G_2\left(\frac{Re}{Re_0}\right) + G_3(\alpha) \tag{5.43}$$

gives the total sound pressure level and

$$St' = \frac{f\delta_p}{U} \tag{5.44}$$

represents the Strouhal number based on the boundary layer thickness on the pressure side. The empirical scaling functions G_1, G_2, G_3 represent the influence of the Strouhal number, Reynolds number, and angle of attack, respectively. The reference values of Strouhal number St'_{peak} and Reynolds number Re_0 are assumed to vary with the angle of attack and are given in [30], p.70, eq. (56) and (59).

Tip Noise. Tip noise is calculated by using the model developed by Brooks and Marcolini [28] which is based on results from trailing-edge noise theory (see above). The sound pressure level is given by

$$L_{p,\mathrm{TIP}} = 10\log_{10}\left(\frac{M^2 M_{\mathrm{tv}}^3 l_{\mathrm{tv}}{}^2 \overline{D}_h}{r^2}\right) - 30.5(\log St'' + 0.3)^2 + 126. \tag{5.45}$$

The Strouhal number is defined as

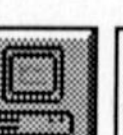

$$St'' = \frac{f l_{\mathrm{tv}}}{U_{\mathrm{tv}}} \tag{5.46}$$

and is based on the extension of the tip vortex l_{tv} and the maximum velocity U_{tv} within or about the separated flow region at the trailing edge. The corresponding Mach number is

$$M_{\mathrm{tv}} = \frac{U_{\mathrm{tv}}}{c_0} \ . \tag{5.47}$$

Blunt-Trailing-Edge Noise. The total sound pressure level caused by a blunt trailing edge is modeled by

$$L_{p,\mathrm{BTE}} = 10\log_{10}\left(\frac{t^* M^{5.5} \Delta s \overline{D}_h}{r^2}\right) + G_4\left(\frac{t^*}{\delta^*_{\mathrm{avg}}}, \Psi_{\mathrm{TE}}\right) + G_5\left(\frac{t^*}{\delta^*_{\mathrm{avg}}}, \Psi_{\mathrm{TE}}, \frac{St'''}{St'''_{\mathrm{peak}}}\right) \tag{5.48}$$

and the Strouhal number based on the trailing edge thickness t^*

$$St''' = \frac{f t^*}{U} \ . \tag{5.49}$$

Comparisons with data for a flat plate revealed a dependence of the trailing-edge angle ψ_{TE} of the airfoil on the sound pressure level. The empirical scaling function G_4 determines the peak level, and function G_5 represents the shape of the spectrum depending on Strouhal number and trailing edge angle. The average displacement thickness

$$\delta^*_{avg} = \frac{\delta^*_p + \delta^*_s}{2} \tag{5.50}$$

is determined from the values of suction and pressure side. The reference value of Strouhal number St'''_{peak} is the ratio $t^*/\delta^*_{\mathrm{avg}}$ and ψ_{TE}, see [30], p.78, eq. (72).

5.3.5 Lowson's Model

Inflow-Turbulence Noise. The model of Amiet [3] was adopted by Lowson [155] for the case of a rotating wind turbines blade [159]. Amiet [3] deduced the formula for an airfoil section under turbulent inflow and validated it against wind tunnel measurements. The fluctuating aerodynamic forces produced by the turbulence in the flow are expressed in equivalent acoustic pressures by using the theories of Kirchhoff [136] and Curle [37]. The spectrum shape is constructed by an approximate formula for a Sears function relating the atmospheric turbulence and the load fluctuations.

The blade is divided into several blade sections along the span. The formula of Amiet can then be applied to every section. Amiet used two frequency regimes, i.e. a low- and a high-frequency regime with the critical frequency $f_c = \beta^2 U/8c$ separating the two regimes, whereas Lowson used an approximate formula which leads to a smooth transition between both regimes. The sound pressure level is given by

$$L_{p,\mathrm{IT}} = L_{p,\mathrm{IT}}^{\mathrm{H}} + 10\log_{10}\left(\frac{K_{\mathrm{lfc}}}{1+K_{\mathrm{lfc}}}\right) \tag{5.51}$$

with

$$L_{p,\mathrm{IT}}^{\mathrm{H}} = 10\log_{10}\left[\rho^2 c_0^2 l_{\mathrm{IT}} \frac{\Delta s}{r^2} M^3 \overline{w}^2 \hat{k}^3 \left(1+\hat{k}^2\right)^{-7/3}\right] + 58.4 \tag{5.52}$$

being the sound pressure level of the high frequency regime. The length scale of the atmospheric turbulence l_{IT} is assumed to be in the order of 100 m.

$$K_{\mathrm{lfc}} = 10 \cdot \mathcal{S}^2 M \frac{\hat{k}^2}{\beta^2} \tag{5.53}$$

is the low-frequency correction factor and

$$\mathcal{S}^2 = \frac{1}{\dfrac{2\pi \cdot \hat{k}}{\beta^2} + \dfrac{1}{1+2.4\hat{k}/\beta^2}} \tag{5.54}$$

an approximate formula for the compressible Sears function suggested by Amiet. The compressibility factor is defined as

$$\beta^2 = 1 - M^2, \tag{5.55}$$

the normalized wave number according to [155]

$$\hat{k} = \frac{\pi f C}{U} \tag{5.56}$$

and the velocity and Mach number

$$U = \sqrt{(\Omega \cdot R_i)^2 + \left(\frac{2}{3} V_w\right)^2}, \quad M = \frac{U}{c_0}. \tag{5.57}$$

 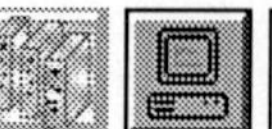 

Trailing-Edge Noise. Lowson uses a re-analysis of the measured data by Brooks, Pope and Marcolini [30]. The peak sound pressure level is scaled using local Mach number M, displacement thickness δ^*, length of blade segment s, and distance to the observer position r. The sound pressure level is given by

$$L_{p,\mathrm{TBLTE}} = 10\log_{10}\left(\frac{\delta^* M^5 s}{r^2}\cdot G_6(f)\right) + 128.5\,, \tag{5.58}$$

the spectrum shape by

$$G_6(f) = \frac{4\left(\frac{f}{f_{\mathrm{peak}}}\right)^{2.5}}{\left(1+\left(\frac{f}{f_{\mathrm{peak}}}\right)^{2.5}\right)^2} \tag{5.59}$$

and the peak frequency by [30]

$$f_{peak} = \frac{0.02\cdot U\cdot M^{-0.6}}{\delta^*}\,. \tag{5.60}$$

Lowson uses the boundary layer thickness of a flat plate to determine the displacement thickness at the trailing edge

$$\frac{\delta}{C} = 0.37\cdot \mathrm{Re}^{-0.2}\,,\ \delta^* = \frac{\delta}{8}\,. \tag{5.61}$$

An empirical factor between 2 and 4 will be multiplied with the displacement thickness in order to account for thicker boundary layers on true airfoils. In a newer version of his code Lowson uses the boundary-layer thickness δ instead of the displacement thickness δ^* for the scaling of trailing-edge noise [131].

Recently Lowson [156] published a refined version of his noise prediction code involving a detailed description of the flow at the trailing edge of an airfoil. The principal idea is to use the formula of Ffowcs Williams and Hall [65] (see Section 3.7.2) and relate it with the boundary-layer flow close to the trailing edge of an airfoil. As documented in [131], Lowson applied a first version to the noise generated by a NACA 0012 airfoil in a wind tunnel. The code is still under development.

5.3.6 Dunbabin

The model of Dunbabin [50] consists of two parts, i.e. (1) an aerodynamic part for the determination of the aerodynamic properties at various blade sections and (2) an acoustic part for the calculation of sound pressures.

The following noise sources are modeled

- inflow turbulence noise,
- trailing-edge noise,
- noise due to separated flow,
- tip noise,
- blunt trailing edge vortex shedding noise,

using parts of the code of Brooks et al. [30]. The inflow turbulence noise is based on the work of Glegg [76] and Grosveld [82].

5.3.7 IAG Model

Like all the previously described models for high-frequency noise, this model (XNOISE) [16] is semi-empirical as well. It is based on work by Brooks et al. [30], Lowson [154] and partly on Grosveld [82]. The acoustic part of the code is coupled with the aerodynamic vortex-lattice method ROVLM (see section 5.2.3). The model incorporates the simulation of the following noise sources:

- trailing-edge noise,
- laminar boundary layer trailing edge interaction noise,
- tip noise,
- noise due to separated / stalled flow,
- blunt trailing edge vortex shedding noise,
- inflow turbulence noise.

The main features of the XNOISE are:

- Models of Grosveld [82], Lowson [154] and Brooks et al. [30] are partly implemented into one code.
- The boundary layer parameters at the trailing edge are determined for real airfoils, i.e. NACA 0012, NACA 4412, FX-79W151A, whereas some of the other models assume the formula for a turbulent flow over a flat plate to be applicable.

- The aerodynamic properties (angle of attack, induced velocities) at the different blade sections are calculated with the vortex-lattice method and passed to the acoustic module.
- An interactive ASCII-based user interface enables a fast and terminal independent handling of the code.

5.3.8 Comparison of Predictions with Experimental Data

In this section, results of current experimental investigations are used to assess the validity and capabilities of the models discussed.

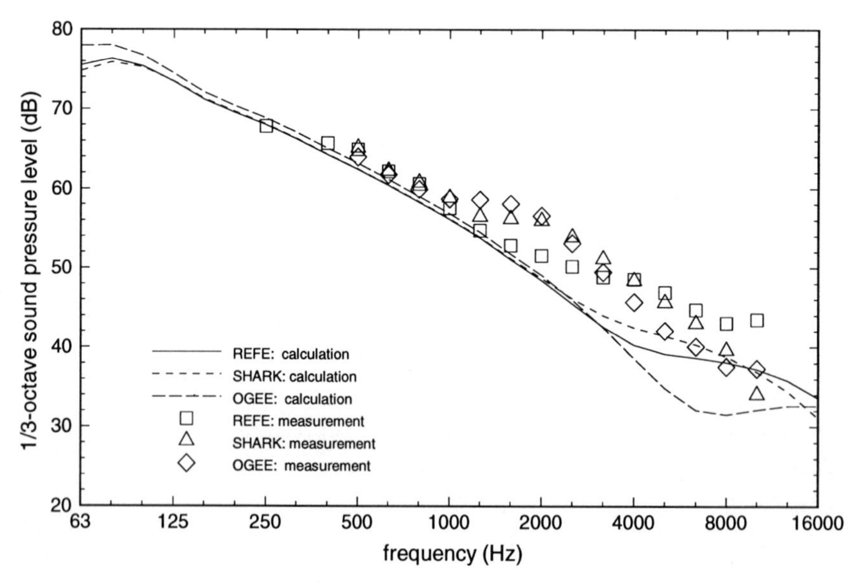

Figure 5.7: Measured and predicted broadband noise spectrum of three blade tips in the wind tunnel [16].

Three different blade tips were measured in the wind tunnel of the University of Oldenburg (see Chapter 7) by DEWI [195] for angles of attack of 4°, 6°, 8°, 10°, 12° and velocities of 20 m/s, 30 m/s, 40 m/s. The noise prediction code of IAG as described in Section 5.3.7 is applied to calculate the broadband sound pressure level generated by the flow around the blade tips. Figure 5.8 shows the predicted and measured spectra for the three tips for an angle of attack of 8° and a wind tunnel speed of 40 m/s yielding a Reynolds number of 562.500. The measured sound pressure levels are corrected for the background noise, and values with a signal/background noise difference less than 6 dB are discarded.

The distribution of angle of attack is calculated with the free-wake code ROVLM (see Section 5.2.2). The acoustic model of Brooks et al. [30] requires boundary-layer parameters close to the trailing edge, which are determined for the applied airfoil, i.e. FX-79-W151A, with the 2D airfoil code XFOIL [48]. Nevertheless, it has to be assumed that the acoustic measurements performed for a NACA 0012 simply scale with the trailing edge boundary layer parameters. This is one of the major drawbacks common for most of the state-of-the-art noise prediction codes that base on the data of Brooks et al. (see Section 5.4).

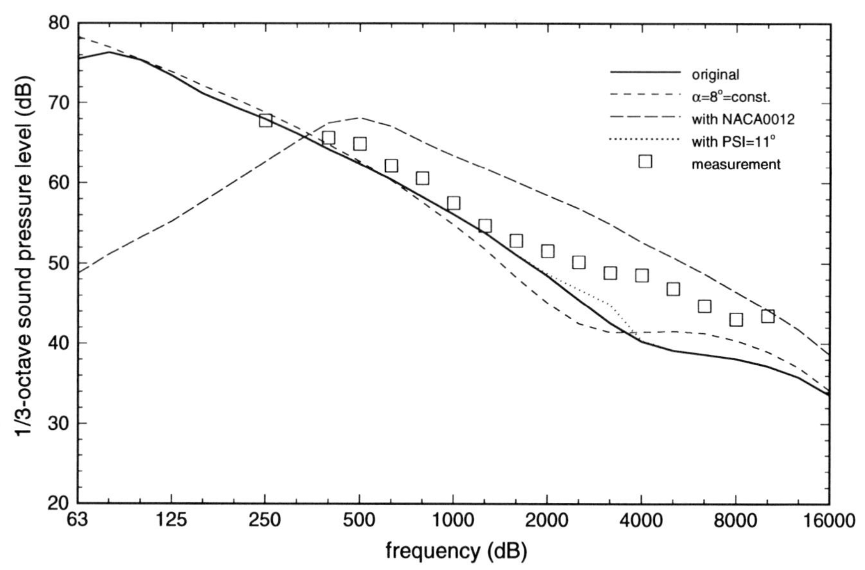

Figure 5.8: Sensitivity of predicted noise to change of input parameters for the reference blade tip [16].

The best agreement between prediction and measurement is achieved for frequencies below 1000 Hz. The agreement is less accurate in the region above 1000 Hz, which is more important w.r.t. annoyance. Although the overall sound pressure level is recovered to a certain degree, the individual differences of the blade tips are not predicted very well. A further discussion of the results will follow in Chapter 8.

A sensitivity analysis (Figure 5.8) shows the influence on noise of the following parameters for the case of the reference tip: airfoil type, observer distance r and aerodynamic model (see below). The original noise spectra are compared to the ones obtained after changing one of the mentioned parameters.

- *Boundary layer parameters* at the trailing edge represent the most sensitive parameter. With the original data used for the NACA 0012 it was not possible to reproduce the measured data, i.e. deviation in the order of 10 dB were found. Only after using the data for the FX-79-W151A, a better agreement was achieved.
- The *aerodynamic model* influences the spectra mainly in the region above 1000 Hz. Originally, the angles of attack for the different blade segments are just set to the geometric angle of attack. After calculation of the angle of attack with the free-wake code ROVLM better agreement between measurement and calculation can be observed in the region 1000–4000 Hz and minor agreement in the region above 4000 Hz.

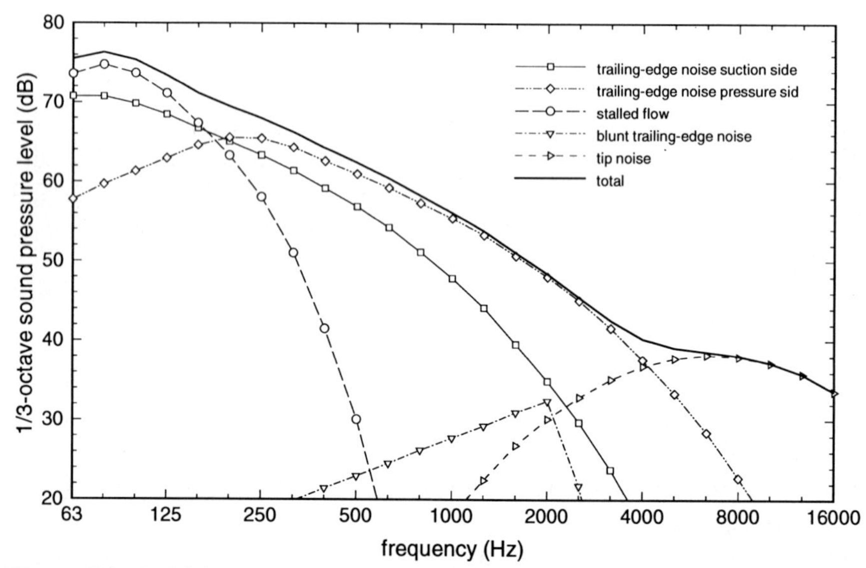

Figure 5.9: Individual contribution of the noise sources to the spectrum of the reference blade tip [16].

Figure 5.9 shows the contribution of the various noise sources to the spectra for the reference blade tip. The total spectrum is dominated by the influence of trailing-edge noise and tip noise.

5.4 Summary

5.4.1 Rules of thumb

Models. The rules of thumb presented are all empirically determined relations between noise level and simple geometrical and/or operational parameters (such as rotor diameter, electrical power output, etc.) [88], [89], [94], [155].

Comparisons. Sample calculations showed that in order to determine an appropriate distance to meet a prescribed sound pressure level, the calculated distances may vary by a factor of two.

Conclusions. This leads to the conclusion that the formulas can only serve as a first guess and are suited for planning purposes only to a very limited extent.

5.4.2 Low-frequency noise

Models. Theory for prediction of low-frequency noise is well-documented in the literature, see e.g. [87], [78], [148]. Prediction models been have been mainly applied to propellers [87] and compressors [150], [151], [152]. In 1981, Viterna [207] successfully applied a code to the low-frequency noise prediction of the MOD-1 wind turbine. Viterna assumed the rotor loads to be concentrated at a representative radius of 75 %. Martinez [164] applied a refined version with radial distribution of forces as input to the same turbine.

Lohmann [146] calculated the sound field of a ducted fan using the solution of the Ffowcs Williams–Hawkings equation and determined the necessary aerodynamic loads with a lifting surface method. In 1993, the free-wake/hybrid-wake code of IAG was coupled with the prediction model of Succi [198] and Farassat [55], [58]. Results and models are described in [16], [83] and [84].

Comparisons. The main difference between the models stems from the methods that are used for the calculation of the aerodynamic loads. The noise prediction models are generally based on the same formulation of the solution for the Ffowcs Willams-Hawkings equation (see Section 3.6.1). The 'older'

 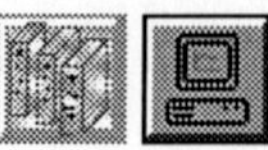 

models mainly use one representative acoustic source at a representative radius, whereas the 'newer' models use a radial source distribution [164] or even a source distribution over the surface [83], [146], [147], [198].

Conclusions. The models generally require only little empirical input, which mainly consists of the velocity deficit around the tower and the airfoil characteristics along the rotor blade. Predictions in the case of a propeller [154], [155] and in the near field, i.e. $r/D < 2$, of a wind turbine [16], are generally of high quality. Existing codes can be applied to solve potential problems. A validation of the codes for a large number of wind turbines would help to minimize future noise impacts caused by low-frequency noise.

5.4.3 High-frequency noise

Models. One of the first prediction models was coded by Grosveld [82] in 1985. A similar model of De Wolf et al. [44] followed in 1986. The model of Glegg et al. [77] is probably the closest to a *Class III* model and should receive more attention in future.

Brooks, Pope and Marcolini [30] performed one of the most extensive investigations with aerodynamic and acoustic measurements conducted for a NACA 0012 airfoil. They scaled their results into a set of easy to implement prediction formulas. A FORTAN77 code listing also given in the report helps to implement the model for a potential user. Nevertheless, the Reynolds numbers Re = 400.000–1.500.000 they used are applicable only to small wind turbines, whereas for large turbines Reynolds numbers up to 5.000.000 could occur.

Dunbabin [49] implemented the code of Brooks et al. and extended it with a model for inflow turbulence. Lowson [154], [155], [157], [158] used the data of Brooks et al. and performed a reanalyzes of their data. He ended up with a simpler model which requires less input and is therefore easier to apply, because input data are often very difficult to obtain. Pettersson [175] implemented a code including models of [82], Brooks et al. [30] and a model allowing the consideration of serrated trailing edges according to Howe [111].

Finally, in [16] codes of Grosveld [82], Brooks et al. [30], and partly Lowson [156] were applied to the case of modified blade tips, which were measured in a wind tunnel [131] and on an experimental wind turbine [24]. Induced velocities were calculated with a free-wake code.

Comparisons. All of these models treat the noise sources separately and use more or less a scaling of the noise spectra of the general form

$$L_p(f) \propto L_{p,\max}(\wp) + G(f) \qquad (5.62)$$

given in (dB). $G(f)$ is a universal spectrum shape function, see e.g. Figure 5.6. The universal spectrum shape function tends towards zero at very low and very high frequencies, with a maximum in between. The maximum sound pressure level $L_{p,\max}(\wp)$ determines the overall level of the maximum of $G(f)$. The parameters $\wp$ usually contain a product of two characteristic lengths and a characteristic velocity.

The noise spectrum for a given case is completely determined after specifying the respective frequency $f_{\max}$ where the maximum sound pressure level occurs. It is defined according to characteristic length scale and velocities of the noise source, e.g. based on trailing-edge thickness t^*, local, effective flow velocity U at the radial position and an estimated Strouhal number St of 0.1 for the case of trailing-edge bluntness noise in the Grosveld model (see Section 3.7.2).

In [16] it was demonstrated that agreement between measurement and prediction can be increased by applying boundary-layer data of the true airfoil (FX79-W-151A) instead of using data for a flat plate or the NACA 0012. The predicted differences of the sound pressure level for three different blade tips were 3–4 dB at an angle of attack of 8° and a flow velocity of 40 m/s compared to differences around 1 dB measured in the wind tunnel and 2–3 dB measured on the turbine. However, operating conditions of the turbines deviate slightly from the wind tunnel case.

Lowson [158] performed a systematic comparison with 22 sets of spectral data from 10 modern turbines and concluded that the mean underprediction is 2.25 dB(A), with a rms error of 1.62 dB(A).

Conclusions. No extensive comparison of different codes for a large number of wind turbines has been available yet. Therefore, conclusive statements concerning the quality of the prediction models are possible only to a limited extent. Investigations by Lowson [158] showed that prediction models could give an estimate of the overall noise of a turbine. Nevertheless, details of the spectral shape are often not reproduced by several codes.

Industry is interested in small modifications of a given blade geometry and its influence on noise. Current models do not contain such a detailed description of the blades and relations to the acoustics. Therefore, models that take the true airfoil geometry into account will be required if design with respect to noise is intended. Work in this direction is underway in EU-financed aeroacoustic projects (DRAW, STENO, see Chapter 9). For the time being, adjustment of the empirical constants in the current models together with measurements could fill this gap until new models are available.

6 Noise Propagation

6.1 Introduction

Chapters 4 and 5 describe noise generation of wind turbines and its prediction, and Chapter 7 deals with noise and flow measurement. This chapter covers noise propagation.

In noise legislation for industrial plants it is common sense to characterize the strength of a sound source by the sound power level L_W, i.e. a measure for the acoustic power radiated away from the source. However, the sound that is perceived by an observer is conveniently characterized by the sound pressure level L_p. In contrast to the sound power level, the sound pressure level is a function of the distance between source and observer. Hence, the sound intensity decreases linearly with increasing distance, see Chapter 2.

In the case of a stand-alone plant, the sound pressure level can be calculated assuming *spherical spreading*, which means that the sound pressure level is reduced by 6 dB per doubling of distance. In the case of a source extending along a line, *cylindrical spreading* has to be assumed, leading to a 3 dB reduction per doubling of distance.

In addition, the effects of atmospheric absorption and the ground effect, both depending on frequency and distance between source and observer, have to be taken into account. Furthermore, the ground effect is a function of the reflection coefficient of the ground and the height of the emission and immission point. The influence of screening or reflecting surfaces is taken into account in a similar way, i.e. the effect is described as a function of screen height and width, and the height of the emission and immission points.

Finally, at distances only slightly above 100 m, weather conditions can have a major influence on the expected noise levels. Figure 6.1 shows the possible change in A-weighted sound pressure level due to weather effects compared to propagation including only spherical spreading and air absorption.

Compared to industrial noise sources, wind turbine noise exhibits some special features. First of all the height of the source is higher by an order of a magnitude, which leads to less importance of noise screening. Moreover, the

wind speed has a strong influence on the generated noise, i.e. $L_p \propto (V_w)^{5\text{-}6}$, including prevailing wind directions, which can cause considerable differences between upwind and downwind positions.

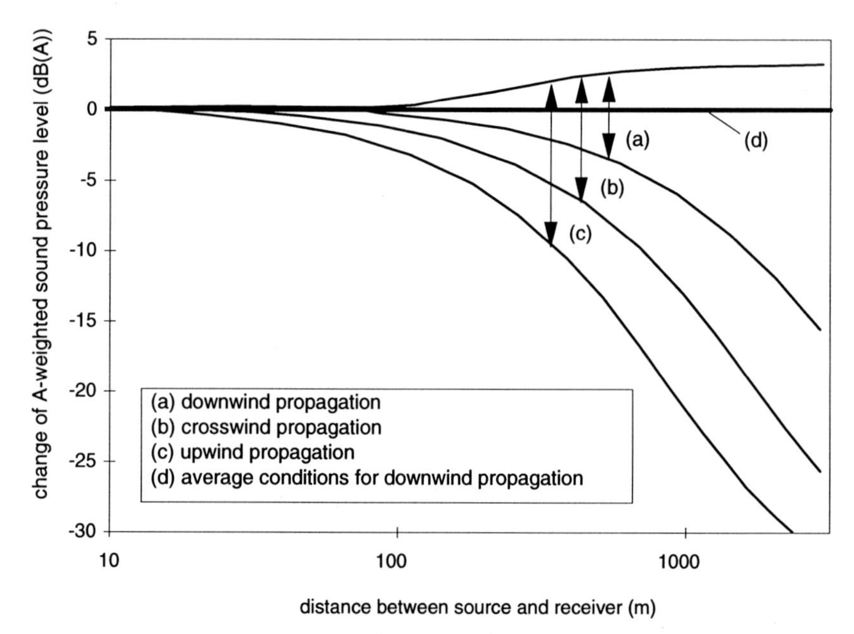

Figure 6.1: Possible change of A-weighted sound pressure level due to weather effects compared to propagation including only spherical spreading and air absorption [205].

Due to the source height, a somewhat smaller spread can be expected for a figure similar to Figure 6.1 in the case of wind turbines, but still a considerable weather influence is foreseen.

This illustrates the importance of an accurate prediction method, which will generally be more precise than a single measurement. The main factors which have to be considered in corresponding codes, can be summarized as

- source characteristics (directivity, height, etc.),
- distance source to observer,
- air absorption,
- ground effect, i.e. reflection of sound at the ground, depending on vegetation, ground properties etc.,
- propagation in complex terrain,
- weather effects, i.e. change of wind speed or temperature with height.

Each of the above mentioned factors is determined separately by using empirical functions, which base on extensive outdoor measurements. Summation of all contributions then yields the total sound pressure level at a prescribed observer position.

Noise propagation codes are used by authorities and planning companies in siting wind turbines / farms, especially for

- simulating the effect of atmospheric conditions and surrounding on noise propagation,
- determining the noise levels from wind turbines in the vicinity, which is a necessary first step in ensuring that noise limits set up by the environmental authorities are kept to,
- selecting an optimal turbine type to be installed.

This chapter starts with a description of the mechanisms responsible for noise propagation. Furthermore, the simple and widely used noise propagation guideline VDI 2714 [205] will be described in some detail, followed by an overview of work previously performed in the framework of an EU-financed project [203] which was directed towards a refined noise propagation model.

6.2 Mechanisms

6.2.1 Air Absorption

Sound propagating through the air is attenuated by viscosity which converts sound energy into heat. The attenuation is dependent on frequency, temperature, pressure, and humidity. The attenuation is proportional to the transmission path length, and may be given as a number of dB per m. The attenuation is negligible at low frequencies, see Table 6.1. Well established tables of values for air absorption coefficients do exist [116], [145], [205].

Table 6.1: Sound attenuation in the atmosphere for a relative humidity of 70 % and a temperature of 10 °C [205].

1/1-octave band center frequency (Hz)	*Sound attenuation* α_L (dB/m)	*Sound pressure level reduction in 1 km distance* (dB)
63	0.000	0
125	0.001	1
250	0.001	1
500	0.002	2
1000	0.004	4
2000	0.008	8
4000	0.021	21
8000	0.052	52

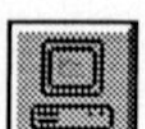

6.2.2 Weather Effects

Both temperature and wind influence the sound pressure level at a given receiver distance. First, the speed of sound c_0 depends on temperature and humidity [86]. Assuming a perfect gas, c_0 is mainly dependent on temperature and increases with increasing temperature, see equation (3.26) in Section 3.3. Second, if sound propagates in the presence of wind, the wind speed component in the direction of propagation will add to the speed of sound. Due to viscous effects, the wind speed is smaller at ground level than higher above the surface.

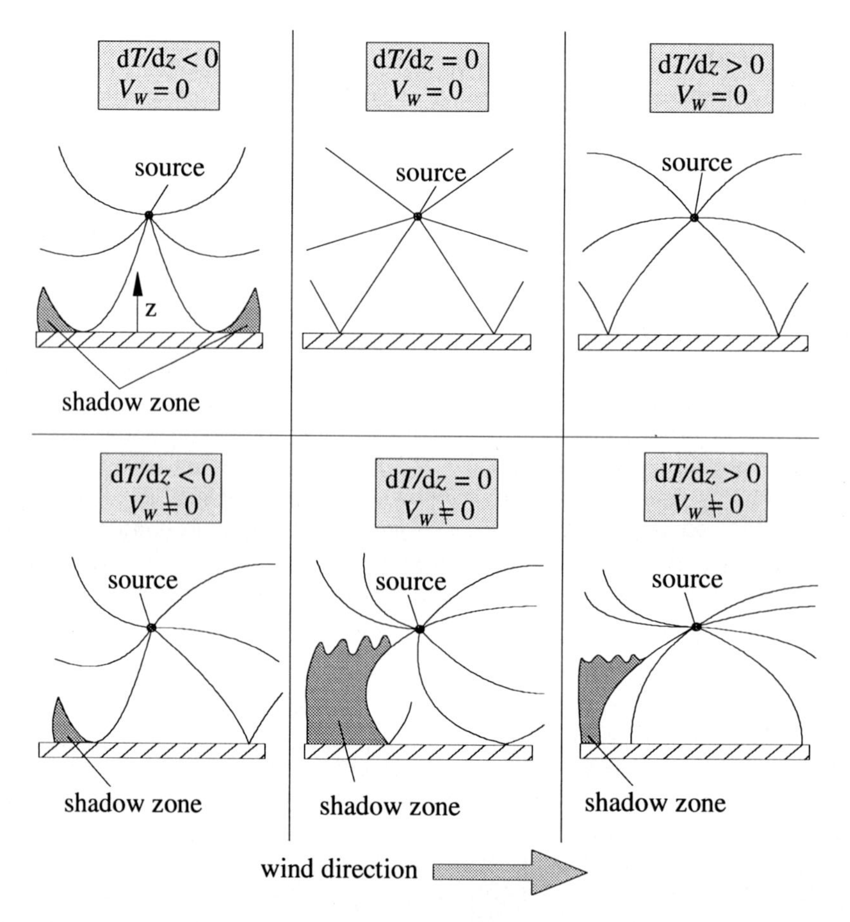

Figure 6.2: Influence of wind speed and temperature gradients on noise propagation [205].

The *wind speed component* is defined as the component of the wind speed vector in the direction of the sound propagation direction. For clarity, sound propagation can be imagined to take place along *sound rays* as is illustrated in Figure 6.2. As for light rays which encounter media with changing speed of light, *refraction* occurs, resulting in curved sound rays.

At daytime, the temperature decreases with increasing height above the ground, i.e. the so-called *lapse conditions*. Thus the sound speed (assuming no wind) will decrease with increasing height, causing the sound rays to curve upwards in all directions around the sound source. At night-time, the temperature in the air layer close to the ground increases with increasing height, i.e. the so-called *temperature inversion*. This causes the sound rays to curve downwards. Typical periods when either lapse or inversion conditions occur are

- lapse conditions: prevailing during clear sunny days, particularly in the summer high altitude of the sun),
- inverse conditions: prevailing during clear nights; in wintertime also possible in daytime.

In the presence of wind in the downwind direction, the sound rays will tend to bend downwards, and in the upwind direction the rays will bend upwards, see Figure 6.2. The effect is that in the downwind direction of a sound source the sound is slightly higher than without wind, and in the upwind direction there is less sound. At a certain distance in the upwind direction a *shadow zone* occurs, characterized by a sudden decrease of the sound pressure level, see Figure 6.2. Based on present calculation models, this is believed to be most pronounced at high frequencies.

It is practical to look at the resulting curvature of the sound rays caused by the combined effects of wind speed gradient and vertical temperature gradient, and to use the radius of curvature as an indicator of the weather influence on sound propagation. It has to be realized, though, that an outdoor sound field cannot be properly modeled by ray theory, and that the actual meteorological conditions are changing continuously. Due to, for example, turbulence, the transmitted sound from a stationary source will fluctuate. Especially shadow zone effects and *ground effect dips* will tend to smooth out, giving less attenuation than predicted on assumption of stationary weather conditions. The *ground effect dip* is used as a term for the pronounced attenuation caused by the ground effect in a region of about 200–800 Hz, see Section 6.2.3.

Refraction of sound waves in the atmosphere causes a shadow zone upwind of the noise source, as illustrated in Figure 6.3. The sound ray just touching the ground surface limits the shadow zone. Kragh and Jakobsen [134] use a formula for determining the edge to the shadow zone which can be applied if

the gradient of the sound speed is constant, leading to constant dT/dz, dU/dz, as well and consequently the sound rays will become a circle. Nevertheless, in reality this is not the case near the ground. According to [118] quoting [86] the radius of this circle can be calculated by means of the following formula

$$\Re = \frac{c_0\left(1+\frac{U}{c_0}\right)^3}{-\frac{U}{c_0}\frac{\sqrt{\kappa R}}{2\sqrt{T}}\cdot\frac{\mathrm{d}T}{\mathrm{d}z}+\frac{\mathrm{d}U}{\mathrm{d}z}} \tag{6.1}$$

for sound rays that are leaving the source vertically. The symbols are

c_0	m/s	speed of sound in air at temperature T,
κ	-	adiabatic exponent for dry air = 1.4,
R	J/(kgK)	gas constant of air = 287 J/(kgK),
T	K	air temperature at reference height,
U	m/s	wind speed component,
dT/dz	K/m	vertical air temperature gradient at reference height,
dU/dz	s^{-1}	vertical gradient of horizontal wind speed component at reference height.

With the symbols given in Figure 6.3, the horizontal distance d from the source to the point where the receiver enters the shadow zone is [134]

$$\begin{aligned} d = d_s + d_i &= \sqrt{\Re^2-(\Re-h_s)^2} + \sqrt{\Re^2-(\Re-h_i)^2} \\ &\cong \left(\sqrt{h_s}+\sqrt{h_i}\right)\sqrt{2\Re} \end{aligned} \tag{6.2}$$

The wind speed component gradient dU/dz based on the logarithmic wind speed profile can be calculated by

$$\begin{aligned} U(z) &= U(10\ \mathrm{m})\cdot\frac{\ln\frac{z}{z_0}}{\ln\frac{10\ \mathrm{m}}{z_0}} \\ \frac{\mathrm{d}U}{\mathrm{d}z} &= \frac{U(10\ \mathrm{m})}{\ln\frac{10\ \mathrm{m}}{z_0}}\cdot\frac{1}{z} \end{aligned} \tag{6.3}$$

using the roughness height z_0. Figure 6.4 shows the distance of the shadow zone as a function of source height and wind speed according to data given by

Shepherd, Grosveld and Stephens [190]. The measurements of Kragh and Jakobsen [134] are included for comparison.

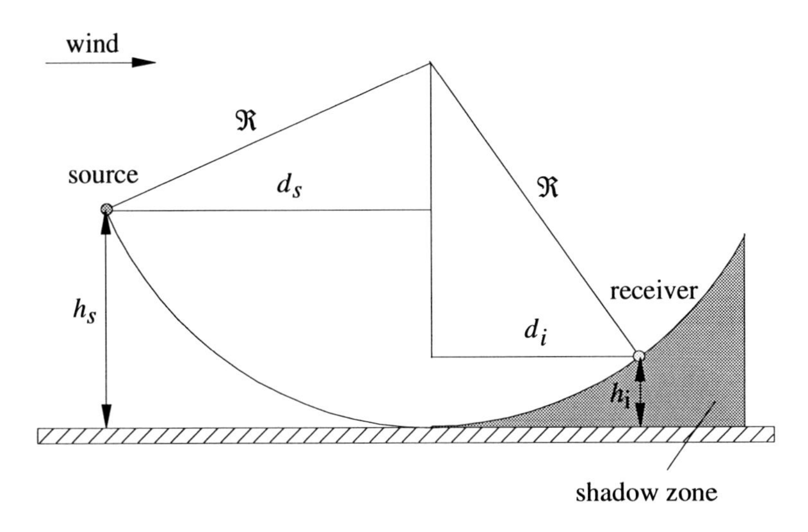

Figure 6.3: Geometrical parameters determining the shadow zone boundary [134].

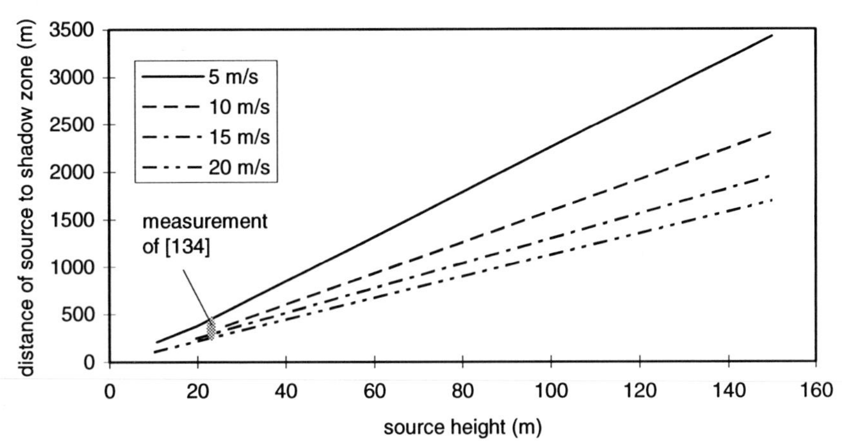

Figure 6.4: Distance to the edge of the shadow zone as a function of source height and wind speed component [190].

Apart from the shadow zone, the main effect of the weather is the modification of ground effect, see next section. The curvature will change the angle of incidence between the reflected sound path and the ground, and

thereby change the "ground effect dip". The more pronounced the ground effect, the stronger is the influence of the weather. In situations with little ground effect (high source and/or receiver, small distance), the weather only slightly influences sound propagation.

6.2.3 Ground Effect

When the sound propagates above a ground surface, the sound ray reaches the receiver without being reflected, i.e. the *direct ray* and the sound ray reflected by the ground will interfere at the reception point. Due to the transmission path difference and the finite impedance of the ground surface, the reflected sound ray will have a different amplitude and phase from the direct sound path, and no complete cancellation will occur. In general, the ground effect will result in a pronounced attenuation of sound in the frequency range 200–800 Hz, and an increase of the sound pressure at low frequencies [134]. At frequencies above 800 Hz, a less pronounced attenuation could be observed. An example of measured and calculated ground effect clearly illustrating the *ground effect dip* is shown in Figure 6.12.

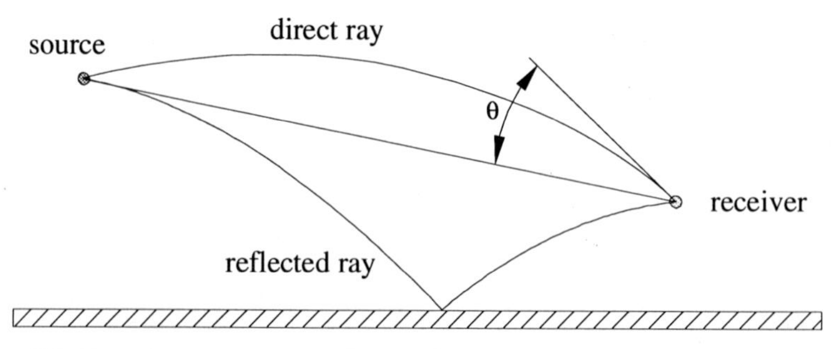

Figure 6.5: Curved ray path with an "arrival" elevation angle θ due to the refraction of sound waves in the atmosphere [134].

Ground effect is the sound pressure level relatively to the sound pressure level in a free field with spherical spreading and absorption in air [134]. The ground effect depends on ground properties (mainly its porosity), geometry (height of source and receiver and the distance between them, defining the angle of incidence with the ground, see Figure 6.5), and on weather conditions (see previous section). The ground effect has been subject to a considerable research effort during the last few years, and the conditions at short distance between source and receiver, where the weather influence is limited, are now well understood. Present work is focusing on sound propagation above complex terrain and at long range.

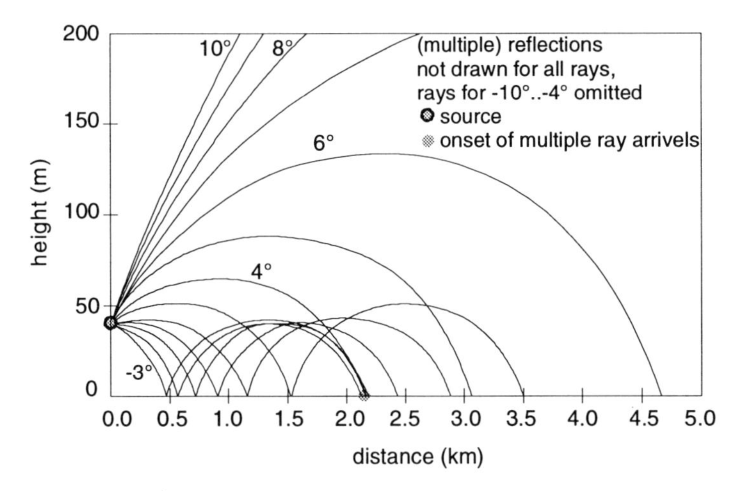

Figure 6.6: Family of sound rays with launch angles ± 10° with an increment of 1°, schematically redrawn from [95].

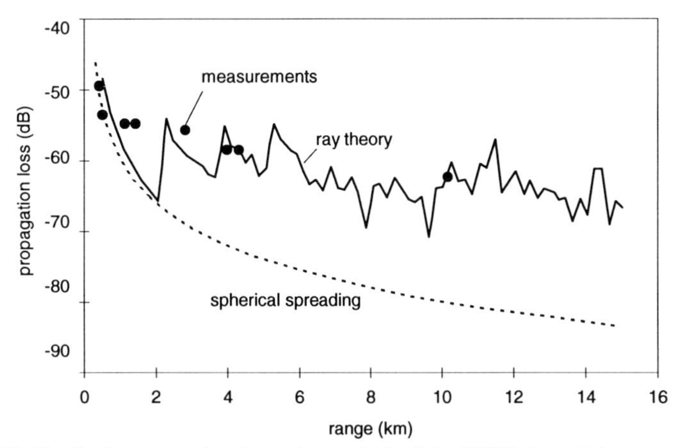

Figure 6.7: Typical propagation loss downwind of the WTS-4 applying ray theory for a frequency of 8 Hz for rays in Figure 6.6 [95].

One example of long range sound propagation of very low frequencies is illustrated in Figures 6.6 and 6.7. Figure 6.6 shows the *sound ray pattern* for downwind conditions, where an increasing number of sound rays reach the ground at increasing distances due to the ray curvature mentioned in the

preceding section. Figure 6.7 shows the calculated *propagation loss* at 8 Hz and compares it with measurements of the low-frequency noise from a large wind turbine. It is found that in the initial 1–2 km the law of spherical spreading is closely followed. The predicted jumps in the propagation loss curve at longer distances are associated with the reception of more sound rays. The measurement results agree reasonably well with the predicted propagation loss.

6.2.4 Screening, Complex Terrain

If an obstacle blocks the line of sight between a sound source and a receiver point, the sound level will be lower. For high frequencies the obstacle creates a shadow zone, i.e. almost no sound can be detected. However, if the frequency is low enough so that the wavelength is comparable to the size of the obstacle, diffraction at the edges of the obstacle will cause that sound waves to be bent into the shadow zone, see Figure 6.8. For very low frequencies with much larger wavelengths compared to the obstacle size no shadow does occur.

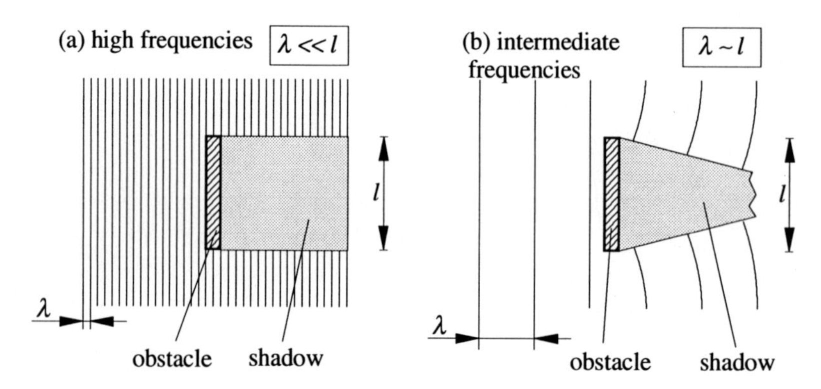

Figure 6.8: Illustration of diffraction of sound waves around an obstacle.

The effect of a screen is not independent of the ground effect, and the weather strongly influences both. In strong downwind, the screening effect of an obstacle may completely disappear, which can be illustrated by the ray curvature causing the sound to pass above the screen.

Most theory on sound propagation has been developed assuming flat ground. This is a necessary first step which must be taken prior to solving the much more complicated problem with sound propagation in a complex (mountainous or hilly) terrain. As a first assumption, the effect of hills that block the line of sight between source and receiver can be regarded as screens.

6.3 Prediction

6.3.1 Overview

From a practical point of view, it is very important to use a prediction technique for noise propagation that is well documented, easy to use and accepted by local authorities. A simple and widely used prediction method for industrial noise is VDI 2714 "Outdoor Noise Propagation" [205] (version of January 1988). It will be described in the following section to illustrate the elements of common propagation prediction methods.

A review of a number of prediction methods for industrial noise has been given by Nolle [170]. Concawe has published a method [35] based mainly on literature that allows for calculation of frequency-dependent corrections for ground effect and weather effects in six different meteorological classes. The general Dutch method has the same basis as Concawe but involves supplementary field measurements of ground effect and screening. It allows for frequency-dependent corrections for air absorption, ground effect, screening, and reflection. It has been implemented with minor changes in the Nordic countries [133] and in Austria, and it forms the main body of the recent international standard for sound propagation calculation [116].

A very common method of this type is the VDI 2714 method "Outdoor Noise Propagation" (version of January 1988), documented in [205]. Due to its importance, it will be described in some detail in the following section. An additional method is described that could become useful in extending the VDI model for ground effects [134].

6.3.2 VDI 2714

The German guideline VDI 2714 is widely used for calculating the effects described in the previous section. It consists of a set of empirical formulas which consider the various effects separately. The sound pressure level L_p at a given observer position is given by

$$\underbrace{L_p}_{\text{receiver}} = \underbrace{[L_W + DI + K_0]}_{\text{source}} - \underbrace{[D_s + \Sigma D]}_{\text{propagation}} \tag{6.4}$$

where L_W is the sound power level, DI is the directivity of the source considering direction dependent sound emission and K_0 considers reflection of the sound in the presence of solid surfaces. K_0 is 0 dB for a sound source located in free field, 3 dB above ground, 6 dB close to two perpendicular surfaces and 9 dB close to three perpendicular surfaces. All units are in (dB). The noise propagation is characterized by the geometrical divergence

$$D_s = 10\log_{10}\frac{4\pi r^2}{1\ \text{m}^2} \approx 20\log_{10}\frac{r}{1\ \text{m}} + 11, \tag{6.5}$$

the distance between source and receiver r, and the sum of all sound reducing effects

$$\Sigma D = D_L + D_{BM} + D_D + D_G + D_e. \tag{6.6}$$

It includes air absorption (see Table 6.1)

$$D_L = \alpha_L \cdot r \tag{6.7}$$

and the ground effect

$$D_{BM} = \left[4.8 - \frac{2h_m}{r}\left(17 + \frac{300}{r}\right)\right] > 0 \tag{6.8}$$

with h_m being the average of source and receiver height. The formula is illustrated in Figure 6.9. With the present definition of ground effect, 3 dB from the K_0 has to be added to the term D_{BM}. The prediction method also allows for calculation of the supplementary reduction due to vegetation D_D, obstacles, and buildings D_G (from multiple reflections and absorption) [205]; this is rarely relevant in the case of wind turbine noise. The effect of noise screens D_e is calculated by use of VDI 2720/1 [206].

Note: The given formulas can only serve as examples. For detailed calculation please refer to VDI 2714 [205].

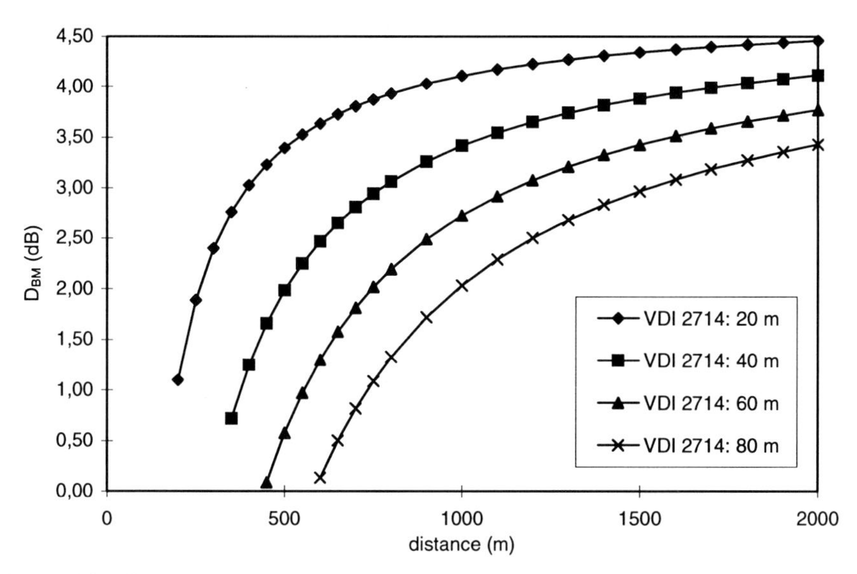

Figure 6.9: Illustration of ground effect for various source heights according to VDI 2714 [205].

6.3.3 DELTA Model

Overview. A more detailed propagation method is outlined in the report by Kragh [134]. Earlier investigations have shown that methods developed to predict propagation of industrial noise are insufficient for predicting the propagation of wind turbine noise [6]. The report describes a series of sound propagation measurements from an elevated source. The measurement results are used as the basis for a method to predict attenuation of wind turbine noise during propagation.

The outline of a detailed prediction method is described. For practical applications, a set of simple formulas for octave band calculations will have to be developed based on the theory. By use of the described principles, a very good agreement is found with the measurements in the downwind and crosswind directions, and an acceptable agreement is established with the upwind (shadow zone) measurement results.

A total of 100 propagation measurements were made, each averaged over 5 minutes. The measurement distances were 100 m, 200 m, and 400 m. The wind speeds were between 2 m/s and 10 m/s and the vertical temperature gradient between 0.14 °C/m and -0.19 °C/m; two thirds of the measurements were performed at night.

The measured sound pressure levels were corrected for spherical spreading and for sound absorption in air according to ISO 9613-1, [116]. Sound power

measurements of the loudspeaker have been made in an anechoic room to determine the source strength. In this way, the ground effect could be determined precisely.

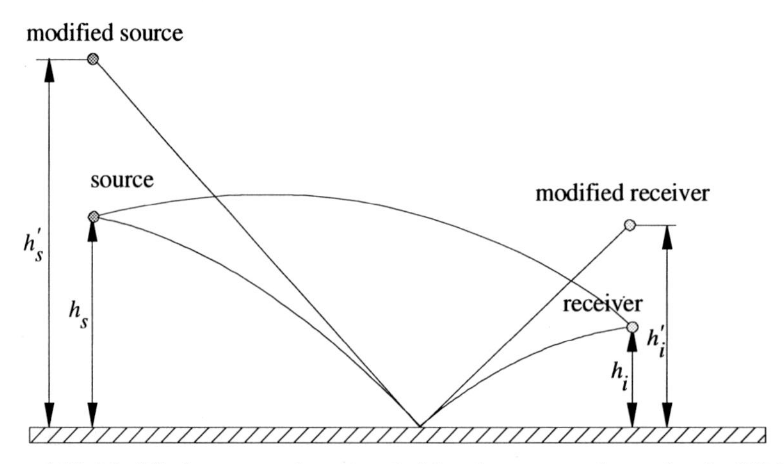

Figure 6.10: Modified source and receiver heights due to curved sound paths [134].

The measured downwind ground effect curves [134] look very much like what was expected to appear for straight-line propagation. Therefore, it was assumed that it is possible to model the measurement results by calculations using theory for straight-line propagation over a finite-impedance ground but using a modified source height, see Figure 6.10 and Figure 6.11. A better approximation might be obtained by modifying both the source and receiver height. Nevertheless, the present model is restricted to this simplification, justified by the fact that the source height of wind turbines is in general much larger than the receiver height.

It is assumed that the downwind refracts the direct sound ray and hence, the parameter is the resulting elevation angle θ of the arriving ray. The elevation angle θ is a function of the distance d between source and receiver and the vertical sound speed gradient. The sound speed gradient is a function of wind speed, terrain roughness etc.

In propagation conditions with shadow zone formation, however, such predictions cannot give the same results as measurements. Therefore, data from the literature on the attenuation of sound propagating into a shadow [190] was implemented, yielding limited agreement of measurement and prediction [134].

In an earlier attempt [133], a more complex propagation model taking the sound speed gradient into account was tested. The results, however, did not look reasonable for large source-receiver distances and consequently were disregarded in the current model.

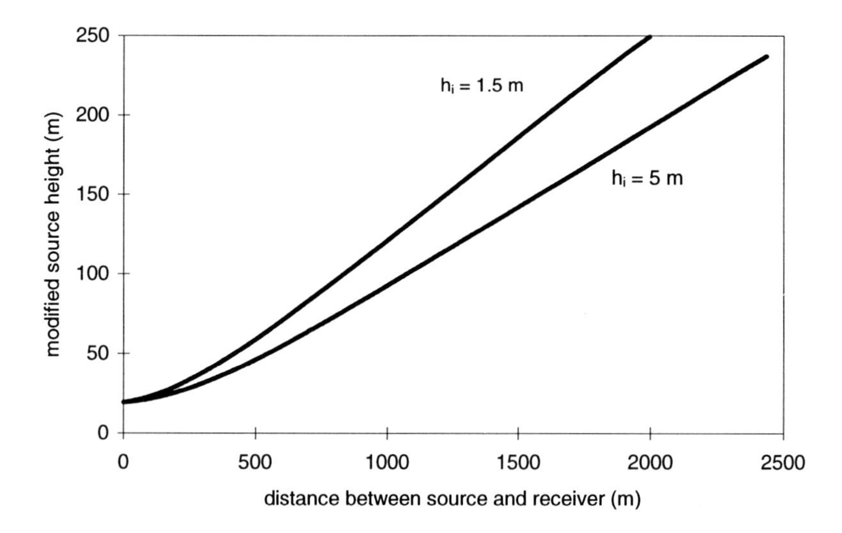

Figure 6.11: Ray-tracing calculations of a modified source height for an actual source height of 20 m to simulate downwind propagation by a straight-line model, h_i is the height of the receiver [134].

Procedure. The possibility of simulating sound propagation along curved transmission paths by straight-line propagation theory has been proposed earlier, for example, in [137] and [129]. The travel time difference between directly transmitted and ground reflected sound is longer with downward curved transmission paths than with straight-line propagation, i.e. downwind corresponds to larger source and/or receiver height.

A ray-tracing program made by DELTA Acoustics & Vibration based on an algorithm described in [135] was used to calculate the elevation angle θ (shown in Figure 6.5) of the sound ray arriving at the receiver. Based on the ray arrival angle θ the modified source height

$$h_s^{'} = h_i + d \cdot \tan\theta \quad (6.9)$$

was calculated. The ground effect is calculated by means of a method developed for straight-line ("no wind") sound propagation by K. B. Rasmussen [181]. This theoretical method is valid at discrete frequencies and therefore a modification for calculations in 1/3 octave bands, taking into account the degree of coherence as proposed in [211], was applied. The

method parameters are distance, source height, receiver height, ground flow resistivity according to the Delany-Bazley impedance model [43], and a coherence factor. The coherence factor was set to 0.1 as suggested in [211] for aircaft noise. The procedure for calculation of ground effect in connection with noise propagation can be summarized as.

Downwind. To calculate the ground effect on noise propagating downwind from a wind turbine the increased height corresponding to the relevant distance, wind speed, terrain roughness etc. has to be determined. This may be done by means of a graph such as Figure 6.11 for the relevant source height or by a ray-trace calculation. Then the ground effect for straight-line propagation has to be calculated.

Crosswind. To calculate the ground effect on wind turbine noise propagating crosswind, use the method for straight-line propagation. The actual source height is used in the calculation.

Upwind. First, calculate the reduced source height corresponding to the relevant distance, wind speed, terrain roughness, etc. Then calculate the ground effect for the straight line propagation using the reduced source height. Finally, calculate the distance to the shadow zone, corresponding to the source height, wind speed and terrain roughness, and calculate the distance of the observation point to the shadow zone and the shadow zone attenuation.

At low frequencies (below the first ground effect dip), the corrections from the straight line model are used. At higher frequencies, if the shadow zone attenuation exceeds the ground effect, this is used as combined ground-and-shadow correction.

In addition to the ground effect, distance (spherical spreading) and air absorption shall also be corrected for.

6.4 Results

6.4.1 VDI 2714

Nolle reports [170] an investigation evaluating the performance of the VDI 2714 (Version 1976) method. Measurements have been performed for 33 different industry sound sources at 75 different receiver positions. The differences to the predictions $\Delta L_p = L_{pcalc} - L_{pmeas}$ obtained with VDI 2714 are given in the following table.

Table 6.2: Differences between measured and predicted sound pressure levels in downwind position using VDI 2714 [170].

Frequency (Hz) / *Differences* (dB)	*63*	*125*	*250*	*500*	*1000*	*2000*	*4000*	L_{pA}
$\Delta L_p = L_{pcalc} - L_{pmeas}$	0.3	0.1	0.7	0.9	0.2	-0.4	0.9	0.5
standard deviation of ΔL_p	2.8	2.8	1.9	1.7	2.4	2.5	3.8	1.4

The predictions were made by using the preliminary draft of the guideline VDI 2714 from 1976. It was concluded that this method is well suited for the calculating the average downwind noise level from industries with many noise sources emitting broadband noise. Unsatisfactory results have been found with single point sources.

6.4.2 DELTA Model

Two sample results of the method reported in [134] are given in this section, i.e. one illustrating the ground effect and another demonstrating the influence of the refined model on a noise propagation calculation for a given turbine.

Figure 6.12 shows the comparison between measured and calculated ground effect at 400 m for a sound source located in a height of 22 m. Each part of the figure shows the average measured ground effect plus and minus the uncertainty u defined as the experimental standard deviation of the mean

$$u = \frac{s}{\sqrt{n}}, \tag{6.10}$$

where s is the experimental standard deviation of the measurement results and n the number of independent measurements in that 1/3-octave band.

A good agreement at all frequencies is found between the measured and the calculated ground effect. When inspecting all measured and calculated ground effect curves (in Appendix 2 of [134]), it is seen that the same good agreement is found in all measurement positions except of 400 m upwind, and 400 m crosswind in the low position.

It can further be observed that the ground effect in a receiver point 5 m above ground is limited to 0 - 3 dB at all frequencies above 315 Hz or 400 Hz irrespective of the wind speed and direction as long as the shadow zone is not entered.

An example of the outlined prediction method is given in Figure 6.13. The relative spectrum of the noise from a typical wind turbine is used as the

starting point, and a source height of 22 m is assumed. The spectrum of the noise in a receiver point 400 m downwind and 1.5 m above ground is calculated. The source spectrum is corrected for air absorption [116] and ground effect, using a wind speed of 6.5 m/s which results in a modified source height of 50 m. The calculation of the straight line ground effect was carried out by using the method of Rasmussen [181].

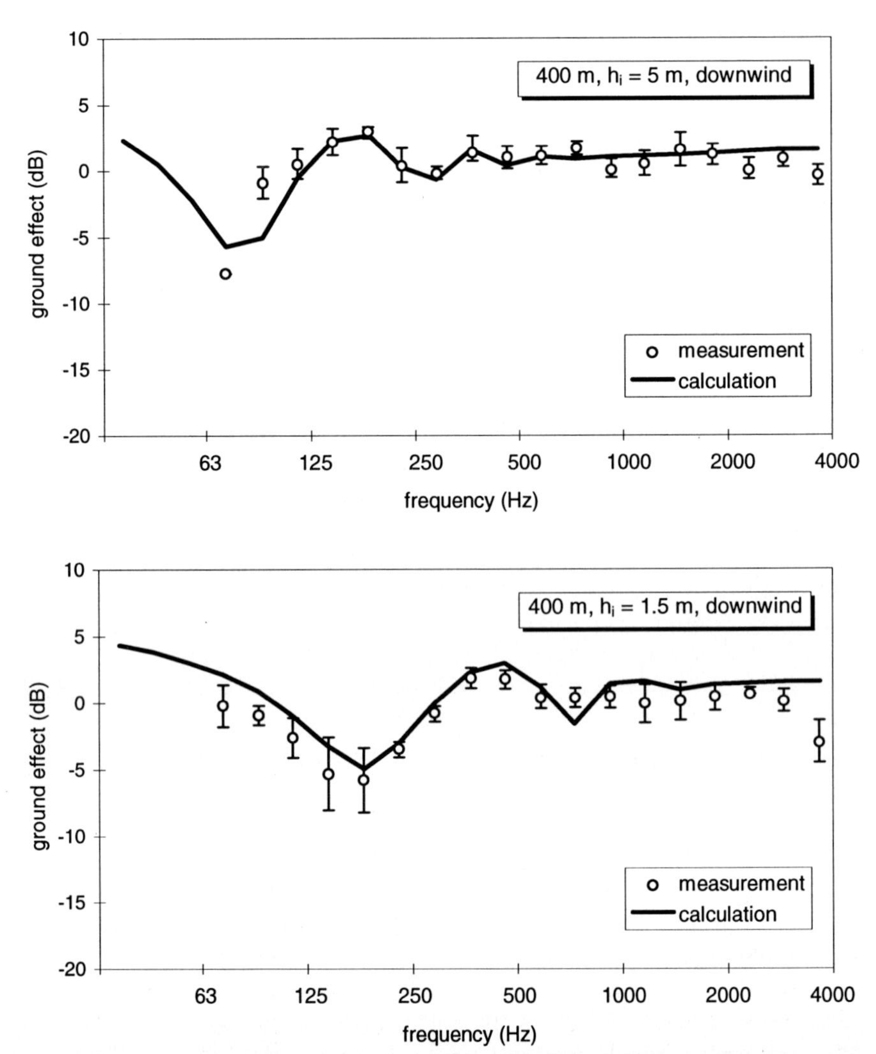

Figure 6.12: Average measured downwind ground effect at 400 m plus and minus uncertainty and calculated values for straight-line propagation using modified source height [134], low receiver position (1.5 m): above, high receiver position (5 m): below.

The measured spectrum is shown, as well. It is based on the average of the measured values of the ground effect from the original investigation for the present geometry and wind speed, where the source spectrum is added and correction for air absorption is also applied. Further the result of a simple calculation, assuming only spherical spreading above a reflecting ground (old method, ground correction + 3 dB) is shown for comparison. The latter has been the official method in Denmark for some years.

It is obvious that the simple procedure overestimates noise levels below 500 Hz and particularly in a region about 250 Hz where the sample wind turbine has a prominent tone. By use of the new prediction method, a 3.6 dB lower A-weighted noise level is found in this example.

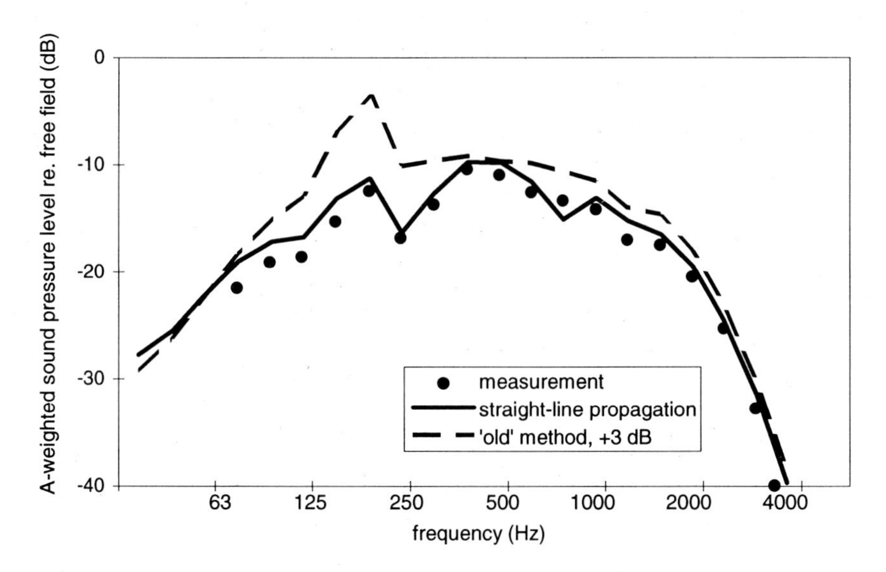

Figure 6.13: Calculated frequency spectra downwind at 400 m, 1.5 m above ground, relative to those in a free field, using measured excess attenuation and the "new" (straight-line) and "old" (+3 dB) prediction method [134].

6.5 Summary

This chapter summarized the mechanisms influencing noise propagation, i.e. air absorption, ground effect, weather effects etc. A widely used noise propagation guideline VDI 2714 [205] was briefly described, followed by an overview of work previously performed in the framework of an EU-financed project [203] which was directed towards a refined noise propagation model.

7 Measurement of Noise and Flow Field

7.1 Acoustic Measurement in the Wind Tunnel

Only recently, wind tunnels have been accepted as facilities for investigating wind turbine noise, although many American airframe noise studies (e.g. [26], [28], [62], [101], [172], [189]) had proved that a wind tunnel set-up offers unique capabilities for airfoil self-noise studies.

The main driving force for the introduction of wind tunnel experiments for studying wind turbine noise was the need to improve the prediction models, and it was realized that this can only be obtained in case the different noise generating mechanism (see Chapter 4) can be studied separately and the local flow parameters affecting these mechanisms can be varied and measured.

In the early nineties, two European facilities were adapted to study the flow-induced noise of sections of wind turbine blades. At the University of Oldenburg in Germany measurements of the noise emanating from different tip sections were performed in the framework of the DEWI project and the ICA project (see Preface) with the aim of explaining the results of measurements performed with different tips mounted on the experimental UNIWEX turbine (see Section 8.4) [24], [131].

In the Netherlands a set-up was constructed in the small anechoic wind tunnel of NLR in the framework of the Dutch national program TWIN. The set-up was defined to allow for measurements of the flow-induced noise emanating from two-dimensional blade sections of 0.25 m chord with serrated trailing edges and of the noise emanating from two-dimensional sections in case the turbulence of the jet flow was varied by mounting different turbulence grids to the tunnel exhaust nozzle [41]. Both set-ups will be described in the following.

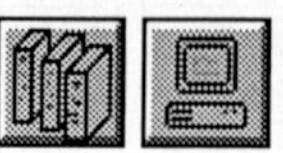

7.1.1 Experimental Set-up at the University of Oldenburg, Germany

Within the framework of the above mentioned projects, wind tunnel experiments have been performed at the Institute for Technical and Applied Physics (ITAP) at the University of Oldenburg [131]. The Oldenburg wind tunnel has an open test section. The nozzle is 1.0 m wide and 0.8 m high. Maximum flow speed is 50 m/s and the degree of turbulence is less than 0.3 % in the center of the stream. For most of the measurements, the velocity range was restricted to 42 m/s. Table 7.1 gives an overview of the technical data.

Table 7.1: Technical data of the wind tunnel at ITAP.

Type	open test section
Nozzle width	1.0 ⊗ 0.8 m
Maximum flow speed	42 m/s (50 m/s)
Maximum chord length of models	0.5 m
Max. Reynolds number based on chord	$1.4 \cdot 10^6$
Orientation of the model	vertical

The acoustic measurements are made with an array of four Sennheiser ME88 directional microphones allowing for signal cross-correlation to subdue part of the tunnel background noise. These waveguide type microphones are placed on the corners of a square with 22 cm edge length. The size was found as a compromise solution between directivity and easy handling. The array is calibrated in a free sound field in an anechoic chamber. The gain of each microphone preamplifier is adjusted to compensate for the slightly different sensitivities of the individual microphones. The microphones are positioned outside of the stream at 1 m distance from the blade's suction side.

All four signals are added up in the time domain and fed into a Hewlett Packard 3569A spectrum analyzer. The useful frequency range of the microphone array used here was found to be 300 Hz to 8 kHz, the transfer function being entirely flat within ≈ 0.5 dB over the whole range. A first order high pass filter in the signal path is used to cut off tunnel noise below 80 Hz, which otherwise would decrease the analyzer's dynamic range.

Different set-ups have been used for the experiments reported in Chapter 8. The results for the three blade tips from the UNIWEX turbine have been obtained with set-up A (see Figure 7.1). No side plates were used. The lower end of the blade was clamped in wooden jaws. This 'vice' was placed on a hydraulic ramp below the stream. The ramp was lowered until the tip was centered in the stream. The larger Danish tips were mounted on a board which was aligned to the lower edge of the outlet nozzle (set-up B).

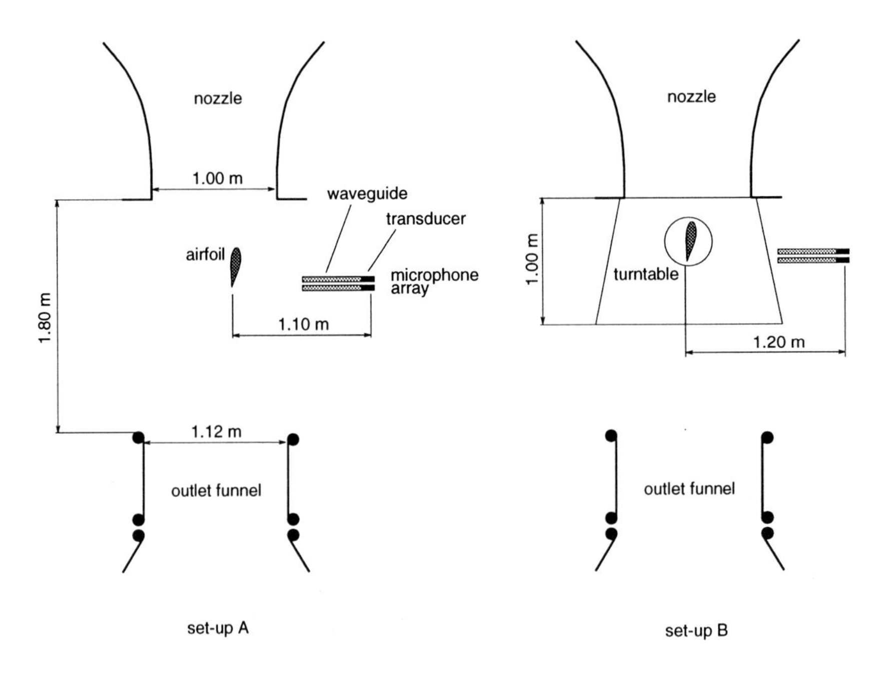

Figure 7.1: Measurement set-up in the wind tunnel of ITAP (top view) [131].

7.1.2 Experimental Set-up at NLR, The Netherlands

Within the framework of the Dutch national program TWIN, wind tunnel experiments have been performed at the National Aerospace Laboratory (NLR) [41]. The maximum flow speed of the $0.4 \otimes 0.6\ m^2$ open jet is 80 m/s, meaning that Reynolds numbers up to approximately $2.8 \cdot 10^6$ can be obtained with models of maximum chord 0.5 m. The hall surrounding the set-up is completely covered with 0.3 m long foam wedges resulting in a cut-off frequency (99% sound energy absorption) of 500 Hz. Table 7.2 gives an overview of the technical data.

After a first series of measurements with the airfoils being placed in a fully open jet, it became clear that more accurate results could be obtained in case interference between the sections (or their support) and the open jet shear-layer could be avoided. Therefore, a semi-open measurement section was realized in which the sections could be positioned between upper and lower end-plates which were mounted to the upper and lower nozzle exhaust lips (see Figure 7.2).

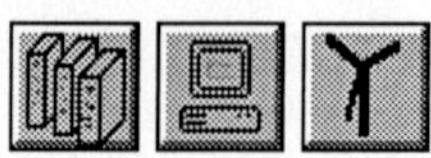

Table 7.2: Technical data of the NLR wind tunnel.

Type	anechoic, open test section
Nozzle width	0.4 ⊗ 0.6 m
Maximum flow speed	80 m/s
Maximum chord length of models	0.5 m
Max. Reynolds number based on chord	$2.8 \cdot 10^6$
Orientation of the model	vertical

Acoustic measurements are performed with an acoustic antenna consisting of 23 microphones and placed 0.5 m aside of the airfoil. The 4 cm spacing of the microphones is based on the localization of the main noise sources up to nearly 6 kHz. The spatial resolution of the antenna enables the separation of airfoil noise and tunnel background noise.

Measurements of the turbulent boundary layer flow along the airfoil surface and of the turbulence generated by the grids is accomplished using hot-wires. By coupling the hot-wires to dynamic data-acquisition and processing equipment, spectra of the turbulence can be measured up to 10 kHz. The set-up further allows for airfoil lift and drag measurements by placing the airfoils on top of a balance. In this way, acoustic performance is related to aerodynamic performance.

7.1.3 Capabilities and Limitations

It is more and more widely recognized that wind tunnel tests can provide unique information about the behavior of airfoil self-noise. In a wind tunnel experiment a large number of parameters can be controlled, adjusted, and/or measured (e.g. flow speed, angle of attack, turbulence level), and be varied in a reasonable short period of time. However, care has to be taken when defining the experimental set-up and when explaining the results and using the findings for full-scale wind turbine applications. The influence of tunnel background noise is discussed below. Scaling effects are discussed in Section 7.1.4.

In contrast to measurements of rotating machinery noise, serious attention has to be paid to the influence of tunnel background noise when performing measurements of flow-induced noise in a wind tunnel. In the case of measurements of the deterministic noise of a propeller or fan, order tracking techniques can be applied, meaning that the signal to background noise ratio can be improved significantly. However, in case non-deterministic noise is generated by the interaction of flow with an aerodynamically well-shaped body like an airfoil, the energy levels of the sound will always be of the same order or even smaller than the energy levels of the tunnel background noise.

Figure 7.2: Measurement setup in the anechoic wind tunnel of NLR.

In reports on wind tunnel measurements of aerodynamic broadband noise, a few methods to reduce the influence of the tunnel background noise are discussed. In the case of the application of single (omni-)directional microphones, an attempt should be made to position the microphones as close as possible to the airfoil and as far as possible from the tunnel background noise sources. In practice this is only possible to a very limited extent since the airfoil has to be placed in the jet potential core, meaning close behind the nozzle from which almost all the tunnel background noise is radiated.

In most recent descriptions on wind tunnel measurements [41], [131] multiple microphone correlation techniques were used to subdue the disturbing influence of tunnel background noise. By cross-correlating the microphone signals, a reduction of the tunnel background noise of 5–10 dB

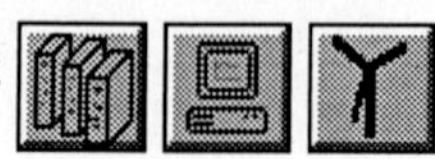

can be obtained [26]. In [24] an acoustic antenna was applied using a 'delay and sum' beam-forming algorithm, and assuming curved wave fronts at the microphone positions. This technique is described in many textbooks (e.g. [100]). It offers relatively high spatial resolution and reduces the influence of the tunnel background noise by more then 10 dB for frequencies above 2 kHz. Acoustic imaging techniques based on near-field acoustic holography have recently been developed [166]. The techniques do not bear the typical limitations of the conventional techniques with respect to resolution and directionality of the source. Some recent publications on the application of these techniques for studying flow-induced noise of cars in wind tunnels indeed report excellent performance [171].

7.1.4 Investigation of Different Noise Sources

Thus far, four different airfoil self-noise mechanisms have been studied in wind tunnel experiments, namely blunt-trailing-edge noise (e.g. [101], [172]), trailing-edge noise (e.g. [26], [62]), inflow-turbulence noise [62], and tip noise [28]. Due to the fact that the details of the flow might have a large influence on the noise generation, it is of the greatest importance to pay attention to the representativeness of the wind tunnel flow. In case full-scale conditions cannot be met, the following scaling-effects have to be taken into account.

The scaling of blunt-trailing-edge noise will not be discussed in this book, first of all since a vast majority of literature is known and many experimental data has been gathered for different applications (a rather complete survey can be found in [22]), and secondly because it is known how this type of noise can be controlled (e.g. by sharpening the trailing edge, see Section 8.6).

In the case boundary-layer trailing-edge noise is studied in a wind tunnel, attention has to be paid to the Reynolds number based on the airfoil chord length. For Reynolds numbers below 1.5 million, whistling tones might occur, due to the partial laminar character of the boundary-layer flow. Occurrence of these disturbing tones can be avoided by gluing roughness strips to the airfoils. At too low Reynolds numbers (approx. < 0.5 million), tripping will lead to turbulent boundary layers which are unrepresentative.

In most of the literature on boundary-layer trailing-edge noise, it is assumed that peak frequencies scale on the ratio of the sizes of the turbulent boundary layer eddies and the convection speed of these eddies. The sizes of the turbulent eddies are assumed to be proportional to the boundary-layer thickness [166] or the displacement thickness [27]. For the convection speed, mostly 0.6 or 0.7 times the free flow velocity is taken.

A recent study on the validity of extrapolating wind tunnel results [156], taking into account the deviating intensities and convection speeds of the turbulent eddies across the boundary layer, has shown that the boundary-layer thickness and a convection speed of 0.7 times the free flow speed are the most appropriate scaling parameters to capture most of the free-field conditions.

One of the earliest reports on the level scaling of trailing-edge noise is given in [66] in which it is shown that the sound radiation of a single turbulent eddy which is convected past a half-infinite plate will be proportional to the fifth power of its convection speed. In the case of a large number of turbulent eddies, the radiated sound power per unit span will be proportional to the fifth power of the velocity, the square of the turbulence intensity and the boundary layer thickness (see Section 3.7).

Where high-frequency inflow-turbulence noise is studied in a wind tunnel, attention has to be paid to the effects caused by the differences in atmospheric turbulence and tunnel turbulence. In a wind tunnel it should first of all be attempted to attain developed, isotropic, and homogeneous turbulence. From measurements on inflow-turbulence noise using grids, as reported in [41], it is shown that grid turbulence, at a typical distance of half a nozzle diameter behind the grid, can be isotropic and homogeneous within approximately 10% of the mean values.

From the literature (e.g. [104]) it is known that the grid turbulence may be considered fully developed only at positions more than hundred mesh widths downstream of the grid. In [41], the distance between grid and airfoil leading edge is reported to vary between 4 and 20 times the mesh widths, leading to a limited reliability of the measurements with the coarsest grid.

In [41] turbulence characteristics and acoustic radiation have been linked assuming that the acoustic frequency scales on the ratio of the size of the turbulent eddy and its convection speed (Taylor hypothesis), which is only true in case the life-time of the eddies is much longer than the time these eddies need to pass the leading edge of the blade. This condition can be met by selecting grids with not too large mesh widths.

Extrapolation to wind turbine cases is made using two dimensionless numbers: the Strouhal number and the Helmholtz number. Here, the Strouhal number is defined as the ratio of the frequency times the chord length and the effective flow velocity, and the Helmholtz number as the ratio of the frequency times the chord and the speed of sound.

The level scaling is addressed in [3]. In case the acoustic wavelength is in between the range of approximately half the chord length and four times the chord length, the radiated sound power per unit span will be proportional to the fifth power of the velocity, the square of the turbulence intensity and the chord length (see Section 3.7.2). At higher frequencies, it is assumed that the dependency on the flow speed and turbulence intensity will still hold but that geometrical parameters have to be included.

A comprehensive experimental study on tip noise carried out in a wind tunnel is described in [28]. From this American research some proof is found for a frequency scaling based on the ratio of the spanwise extent at the trailing

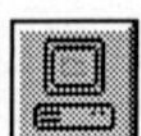

edge of the tip vortex separation region and the maximum velocity along a separation angle, and for both parameters, empirical prediction formulas using the effective angle of attack at the tip and the free flow velocity, are given.

Since there is a definite lack of these data for divergent blade tips, these scaling laws will be of limited value. For this reason, a scaling law proposed for the prediction of the level of tip vortex formation noise, using similar parameters, is not discussed in this section.

7.2 Acoustic Measurements on Operating Turbines

7.2.1 Ground Board

The standard procedure for the measurement of the total acoustic sound power level of a wind turbine has been defined in the IEA-recommended practices for noise measurements on wind turbines [145] and in the draft IEC-standard for noise measurements on wind turbines [114]. Both procedures are based on the use of a measurement microphone positioned on a hard board, placed on the ground at prescribed measurement positions. The use of a ground board diminishes the noise induced by the wind in the microphone and the influence of the ground reflection is made independent of the test site.

The ground board is flat and acoustically hard, e.g. plywood with a thickness of at least 12 mm. The minimum length, width or diameter of the board is 1 m. The measurement microphone has a maximum diameter of 1/2 inch (~13 mm) and is mounted horizontally on the board, facing the turbine tower. The microphone is covered with the half of a normal wind screen. More details are given in Figure 7.3 and Figure 7.4.

The free field sound pressure level spectrum at the measurement position is obtained by subtracting 6 dB from the spectrum measured on the ground board for all frequencies. Note, that the ground board technique does not allow to distinguish between the noise from the nacelle and from the rotor. The method described in [145] gives the apparent A-weighted sound power level of the turbine, its dependence on the wind speed, and the directivity distribution of the noise which is emitted by the turbine.

In order to determine the sound power level as a function of wind speed, the ground board is placed at the reference measuring point 1 which is located downwind the turbine at the reference distance R_0 (see Figure 7.5). If necessary, the board should be inclined in such a way that the angle γ is between 30° and 45°. The sound pressure level is recorded together with the produced electrical power output. The wind speed V_h at the hub height h is deduced from the power curve of the individual turbine.

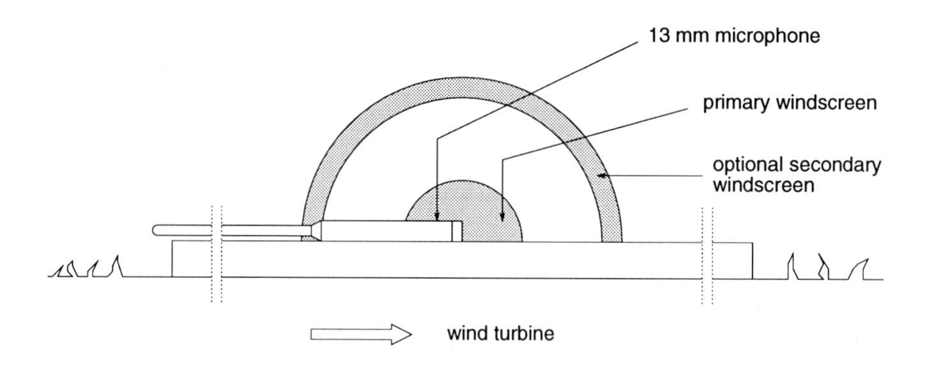

Figure 7.3: Mounting of the microphone on ground board; vertical cut [145].

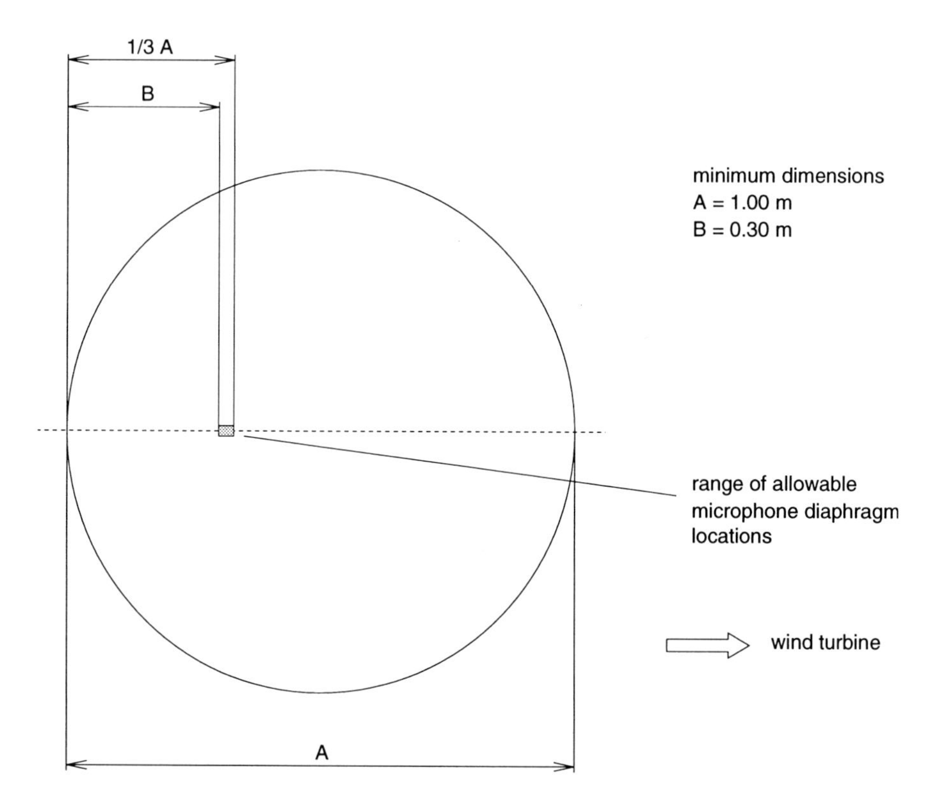

Figure 7.4: Mounting of the microphone on ground board; top view [145].

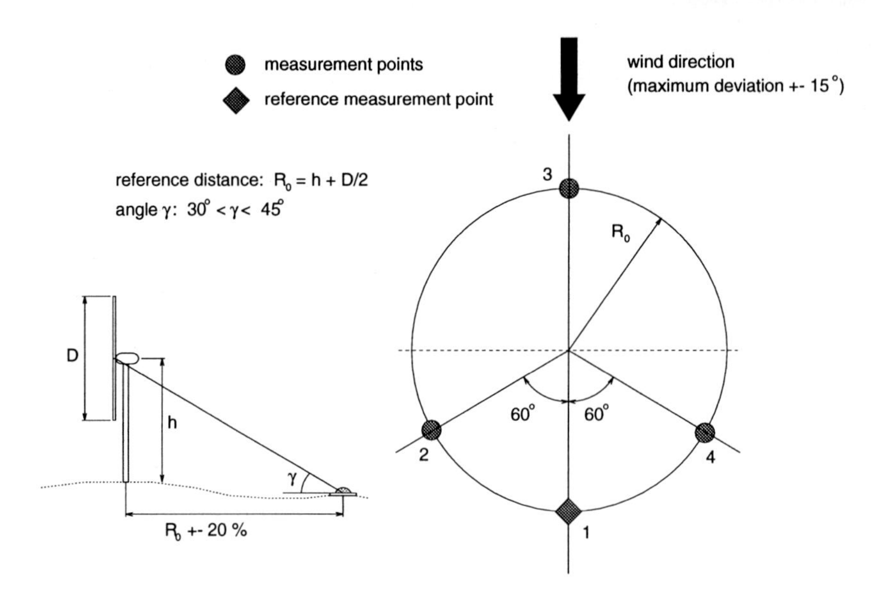

Figure 7.5: Recommended pattern for measurement points; definition of reference distance R_0 and angle γ [145].

This procedure makes the noise measurement independent of test site [130]. The wind speed V_{10} at the reference height 10 m is calculated with the power law and a roughness height of $z_0 = 0.05$ m

$$V_{10} = V_h \cdot \frac{\ln(10/z_0)}{\ln(h/z_0)}. \tag{7.1}$$

A regression line through the recorded data points is determined and the sound level at the reference wind speed of 8 m/s is read. The additional measuring points 2–4 are used for the determination of the directivity distribution. For more details see [145].

7.2.2 Acoustic Parabola

An acoustic parabola is a noise reflecting and focusing structure in a parabolic shape. A normal measurement microphone is mounted in the focal point of the parabola. The parabola causes an amplification of the noise at the position of the microphone. The amplification of the sound pressure in the focal point depends on the angle of incidence of the sound on the parabola and on the ratio between the wavelength of the sound and the diameter of the parabola.

The parabola is positioned on the ground in upwind or downwind direction from the turbine. By aiming the parabola at a certain spot in the rotor area, the sound from this spot will be amplified and the sound outside this spot will be amplified less or not at all. In consequence, the parabola acts as a geometrical filter with a certain spatial resolution. The spatial resolution is low for a small ratio between the parabola diameter and the sound wavelength (thus for low frequencies and a small parabola) and high for a large ratio (thus for high frequencies and a large parabola). The amplification as a function of sound incidence angle and the spatial resolution as a function of frequency are quantified by calibration for a given parabola. This can be done by mounting a loud speaker, emitting broad band noise, temporarily on the nacelle of the (parked) turbine. At the measurement location the parabola and a ground board set-up (see section above) are positioned next to each other.

With the parabola aimed at the loudspeaker, the microphone signals from the parabola and the ground board are recorded simultaneously. The spectrum from the ground board is corrected for the 6 dB to arrive at the free field sound pressure (see Section 7.2.1). The difference between the spectrum from the parabola and this free field spectrum is the amplification. The calibration of the spatial resolution is done by repeating this procedure a number of times while the parabola is aimed at certain distances from the loud speaker. The position that results in a reduction of the amplification of 3 dB corresponds to a certain angle of incidence. This (frequency dependent) angle of incidence is called the sensitive spatial angle.

The parabola can be used for measuring parts of the rotor. Since the rotor rotates and the parabola is stationary, the parabola "sees" the blades of the rotor passing by during the rotor revolutions. For this reason, the sound signal from the parabola is analyzed in, for example, 36 consecutive time intervals with a length of 1/36 of the rotor revolution time (time window technique). Selecting the two (2-bladed rotor) or three (3-bladed rotor) time intervals corresponding with the passage of the blades, the noise from the various blades can be distinguished.

To increase the total measurement time per blade, the results of many rotor revolutions are averaged. This is done by triggering the analyzing equipment once per rotor revolution using an electronic switching device on the rotor shaft. The disadvantage of the parabola is the limited resolution at low frequencies: a parabola with a diameter of 1.8 m at a standard distance from the turbine (hub height plus the half of the rotor diameter) cannot distinguish the blades from the nacelle for frequencies lower than about 500 Hz. The advantage of the parabola is that at higher frequencies the noise production of different blades can be measured at virtually the same moment (under identical conditions) by mounting the blades on the same rotor (see Section 8.4.2).

Relative measurements. With a parabola it is easy to perform relative measurements. For example, the parabola can be aimed at the position where

the blade tips are going in a downwards direction. The measured spectra of the different blades on the rotor are subtracted by the spectrum of one of the blades (the reference blade). This requires no calibration of the parabola.

Absolute measurements. A parabola can be used to measure the noise production of the different blades on the rotor as a function of the radial sections. By summation of the contributions of the various radial sections, the total noise production of the different blades can be determined. Absolute measurements require parabola calibration as described before. Special attention has to be paid to the selection of the measured radial sections. The sections must be chosen in such a way that the midpoints of the sections (the aiming positions of the parabola) are separated from each other by about twice the value of the sensitive spatial angle. Too large steps would result in missing certain sections and too small steps would result in multiple counting of certain sections. Since the sensitive spatial angle is dependent on the frequency, the aiming positions are dictated by the highest frequency of interest.

During the data treatment, the results of all aiming positions are used for the higher frequencies and a selection of the aiming positions is made for the lower frequencies.

7.2.3 Proximity Microphone

A proximity microphone is a measurement microphone positioned in the proximity of the rotor position to be measured. In principle, the microphone can be co-rotating with the blade but this leads to a high flow speed at the microphone and consequently to a high level of induced noise. For this reason, proximity microphones are stationary.

Using the time window technique (see Section 7.2.2), the signal from the microphone is analyzed for the moments that the blades are close to the microphone. The spatial resolution of the microphone is obtained by the geometrical attenuation of noise from a point source, i.e. 6 dB per doubling of the distance. For this reason, the best resolution is obtained with the smallest distance between the microphone and the blade as can practically be reached. In contrast with the parabola, this spatial resolution is not dependent on the frequency.

A disadvantage of a proximity microphone is the unknown wind-induced noise, especially if the microphone is mounted downwind of the passing blades. Another disadvantage is the fact that the microphone is measuring close to the source (the near field). Therefore, a prediction of the far field noise is impossible. For this reason, the proximity microphone can only be used for comparative measurements of different blades on the same rotor.

7.2.4 Acoustic Antenna

An acoustic antenna is a row of measurement microphones at certain distances from each other. The signals from the different microphones are analyzed by correlation techniques, which results in a spatial resolution of the antenna. The microphones can be positioned on the ground but it is also possible to use a row of proximity microphones (see above). The advantage of using proximity microphones is that the directivity of the antenna is translated to a higher spatial resolution due to the shorter measuring distance. The disadvantages, however, are the one mentioned already in Section 7.2.3. Acoustic focusing is obtained by correlating the microphone signals in the time domain or in the frequency domain. In its most straightforward application it is assumed that planar waves reach the microphones which means that the antenna has to be placed at a fairly large distance from the wind turbine. In practice, a distance of at least one time the hub height is used.

The resolving power of an antenna increases linearly with the frequency and is determined by the length of the antenna relative to the acoustic wave length. For example, an angular resolution of 1 degree (based on 3 dB attenuation) for a frequency of 1 kHz requires an antenna with a length of 20 m. The highest frequency that can be measured is determined by the microphone spacing. Depending on the antenna view angle, a spacing of one to two microphones per wavelength has to be used to avoid spatial aliasing. Therefore the range of frequencies is normally determined by the number of microphones available.

7.2.5 Uncertainties of Acoustic Outdoor Measurements

Two possible effects have to be considered when acoustic measurements on operating turbines are performed:

- turbines of equal make and model may have different acoustic characteristics,
- the acoustic characteristics of a turbine may change from day to day (repeatability) or during the years (aging).

Both effects add an additional uncertainty to the results apart from the usual measurement uncertainty. The ECN project (see Preface) aimed at the quantification of these uncertainties [203]. Information has been collected on the individual differences of sound power level and tonality of turbines of equal make and model by measuring 6 different types of turbines. Of each type, 5 individual turbines have been measured.

In addition, measurements have been performed to quantify the day-to-day variations of the sound power level and tonality of three different turbines. Finally, the effect of aging on sound power level and tonality has been

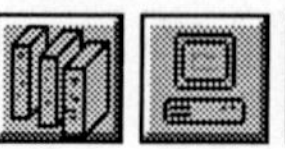

investigated by the repeated measurements of 5 wind turbines that were measured in an identical situation 3–6 years earlier. Sound power levels have been determined in compliance with the recommendations in [145] (see Section 7.2.1). For the determination of tonality, an undisturbed measurement period of a few minutes has been selected at a wind speed as close as possible to 8 m/s, defined at a height of 10 m. The sound pressure signals have been analyzed into narrow-band spectra.

The uncertainty in sound power level, expressed as one standard deviation, due to individual differences of otherwise identical turbines amounts to 0.5 to 1.6 dB(A). The overall uncertainty is 1.2 dB(A). The uncertainties in tonality are very high in comparison with the tonality values themselves. From the results it can be seen that the measured tonality of one turbine gives no reliable information about the tonality of another turbine of equal make.

The uncertainty in sound power level due to reproducibility (day-to-day variations) is equal to 0.3 dB(A). This uncertainty is again expressed as one standard deviation. Given this low value, it can be concluded that this value reflects the measurement uncertainty and that the sound power levels show no day-to-day variations. However, the tonality of a turbine measured on one day gives no reliable information on the tonality of the same turbine measured on another day. The results gained from repeating noise measurements on five turbines after an interval of 3–6 years have shown a tendency towards increasing the A-weighted noise level by up to 2 dB. The increase is small compared to the expected measurement uncertainty. The limited number of observations, however, does not allow for a definitive conclusion on the significance of the increase.

7.3 Flow Visualization on Operating Turbines

7.3.1 Oil–Soot

The flow visualization with a mixture of oil and soot, which is well known from wind tunnel experiments, is a rather simple yet efficient method of surface flow visualization. An oil–soot mixture is painted on the blade in the region of interest. The measurement has to be started immediately in order to achieve steady flow conditions as fast as possible. After a few minutes the mixture will be dry and can be photographed. The oil–soot mixture has minimum adverse aerodynamic effects.

Apart from the generation of a two-dimensional streamline picture, this flow visualization method allows the determination of the transition point between laminar and turbulent flow, the regions of flow separation, and the vortex generation at the surface.

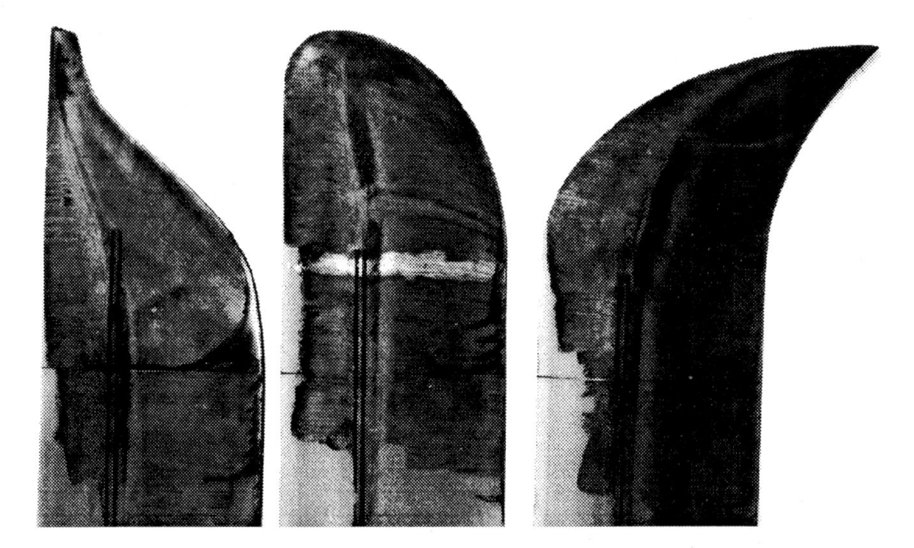

Figure 7.6: Example for flow visualization with oil-soot for the blade tips discussed in Section 8.4.2; suction side [2].

Important is the right mixture of oil and soot. Too little oil causes the mixture to dry too fast, whereas too little soot does not lead to a useful streamline picture. This relation depends on the measurement conditions and has to be found out by trial and error. Also, the right viscosity of the mixture has to be found, depending on the desired wind speed. In practice, petroleum has shown good properties for this kind of flow visualization for a blade velocity of around 50 m/s.

In the case of a rotating wind turbine which cannot be driven with an artificially constant wind speed it is obvious that for each blade tip to be investigated several measurements have to be performed. In case of aeroacoustic investigations, normally the suction side is of interest. In some cases also the pressure side of the blade could be important. In order to get some information about the tip vortex behavior, both the upper and lower side of the blade tip region should be included in the visualization. Figure 7.6 shows an example for flow visualization in the wind tunnel for the tips used on the UNIWEX turbine [2] (see Section 8.4.2).

Ordinary painting equipment like a foam material roller or a spray gun can be used to put the oil–soot mixture on the blades. Especially in the case of large wind turbines, the painting of the blades can cause severe difficulties due to the accessibility of the blade tips. Similar problems arise for photographing the results.

 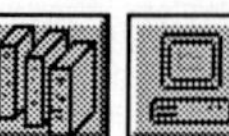 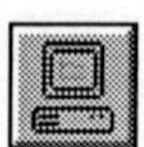 

The interpretation of the oil–soot pictures is difficult since a rotating system is investigated. Due to the centrifugal forces on the oil, which has a much higher density than the air, the streamlines on the oil soot pictures can show a different behavior than the streamlines of the actual flow. Despite this problem, however, the regions of separation and those where the tip vortex "wets" the surface can be detected as in the case of a non-rotating system.

During the flow visualization campaign of one blade tip, it is not necessary to clean the blade tip properly, because it is possible to paint over an old oil–soot picture several times. After the measurements, the oil–soot mixture on the wing can be easily removed with a household window cleaner, provided that the blade surface is polished or lacquered.

7.3.2 Tufts and Co-Rotating Video Camera

The surface flow visualization method mainly used is tufts. This is a light flexible material that is attached to the surface and aligns itself with the local surface flow. The streamline pattern can be recorded by using photographic methods or by video. Apart from the streamlines, the area of separation and vortex surface interaction can be visible on the tuft recordings.

The tufts are normally polyester or cotton sewing threads which can be photographed either in ultraviolet or white light. In contrast to the oil–soot flow visualization method, the tufts do affect the flow properties such as the aerodynamic loads. Acting like vortex generators, they can cause transition from laminar to turbulent flow already close to the leading edge. The aerodynamic influence of the tufts can be minimized using so-called 'minitufts'. In this case, the tuft material is monofilament nylon (diameter $\approx$0.04 mm) which has been treated with a fluorescent dye and must be photographed in ultraviolet light. The centrifugal forces on the tufts are small in relation to the aerodynamic forces and can be neglected in the interpretation of the streamlines.

Using tufts for surface-flow visualization is a common practice for experiments in wind tunnels and with a fixed experimental set-up. Experiments are quite new with rotating systems like wind turbines. In co-operation between CARDC (China Aerodynamics Research and Development Center) and FFA (Aeronautical Research Institute of Sweden), flow-visualization tests on a 5.35 m diameter wind turbine were carried out in the CARDC's large low-speed tunnel [46]. Since the blade is rotating, the recording system has to be more complex. The easiest way is a rotating camera, mounted on the hub or on the blade near to the region of interest. For the hub mounted case a high zoom lens is necessary, while it has to be assured in the blade mounted case, that the flow in the investigated region is not disturbed and one can handle the centrifugal forces.

Figure 7.7: Example for flow visualization with tufts and co-rotating video camera on the UNIWEX turbine [24] (see Section 8.4.2).

In both cases the camera focuses the blade tips with constant distance and direction, so the zoom can stay fixed. Due to the rotating system quite different brightness conditions occur, because the picture background varies between the ground and the sky during one revolution. If the camera is not able to adapt to the varying brightness conditions, the exposure time of the camera has to stay constant and only a certain section of one rotor revolution can be used for later evaluation. Figure 7.7 shows two examples for the use of tufts on the UNIWEX turbine [24].

In order to obtain constant brightness conditions the measurement could be undertaken during the night-time, with tips illuminated by co-rotating spots lights. Anyhow, vibration and blade movement under operational conditions must be taken into account.

Another possibility is to position the camera on the ground at a certain upwind or downwind distance and to rotate it in such a way that it follows the rotating tip. This rotation has to be synchronized with the rotor of the turbine. Since the distance to the blade tip varies, because the camera is not mounted in hub height, it is likely, that the focus has to adapt fast during one revolution. Again, a high zoom lens of good quality is needed. Such a set-up leads in most cases to a distinctly greater viewing angle than a configuration with a hub mounted camera and, therefore, allows a better evaluation and resolution of the flow field.

8 Noise Reduction

8.1 Introduction

Chapter 5 has described the approaches for predicting the noise emission due to the different noise mechanisms introduced in Chapter 4. It became clear that current prediction codes can give an overall estimate of the noise of a blade section or a whole wind turbine, but often fail to predict correct spectral shapes. However, since the mechanisms of noise generation are partly well understood there will be a possibility of deducing ways to *reduce* the noise even if it is not possible to *compute* it correctly. Based on theoretical findings and extensive experimental work, this chapter shows how a noise reduction may be accomplished.

Research work performed in the past [32], [41] showed that three mechanisms of noise generation are important for wind turbines, assuming that all tonal contributions (due to slits, holes, trailing-edge bluntness, etc.) can be avoided by proper blade design:

- trailing-edge noise,
- tip noise,
- inflow-turbulence noise.

This chapter starts with some simple estimates, how much noise reduction can be achieved by changing tip speed and pitch setting in an appropriate manner. The main part focuses on the reduction of trailing-edge noise, tip noise, and inflow-turbulence noise. Finally, some experimental results concerning blunt-trailing-edge noise are reported.

The effect of blade modifications (tip shape, trailing-edge shape) and operation (tip speed, pitch) on noise has been investigated within the various EU projects (see Preface). These investigations included wind tunnel experiments in two acoustic wind tunnels (see Section 7.1) and outdoor measurements on several commercial and experimental wind turbines. The most important results of these projects are presented in this chapter. For a

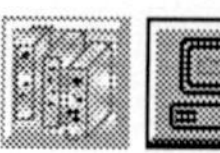

detailed description of the work see the respective final reports [24], [131], [203]. The experimental set-ups in the two acoustic wind tunnels and the procedures used for acoustic outdoor measurements are described in Chapter 7.

8.2 Reduction of Tip Speed and Angle of Attack

The different means for noise reduction described in this chapter partly require considerable changes in the geometry of the blades, especially of the tips and the trailing edges. However, there are much easier ways to reduce the noise by using the strong dependence on the flow velocity at the tips and on the pitch setting.

Theoretical analysis [3], [23], [66] revealed that both trailing-edge noise and inflow-turbulence noise, which are the dominating sources, change with M^5 (see Section 3.7.2). Here M is the Mach number which is highest at the blade tips. The flow speed at the tip depends on the rotor frequency Ω, the blade diameter D, and the wind speed V_w

$$V_{\text{tip}} = \sqrt{\left(\Omega\frac{D}{2}\right)^2 + V_w^2} = \Omega\frac{D}{2}\sqrt{1+\frac{1}{\lambda}} \tag{8.1}$$

where $\lambda = \Omega D/2V_w$ is called the tip speed ratio. The formula by Hagg [89] (see Section 5.1.3) allows to estimate the reduction of A-weighted sound power level in case the rotational speed or the rotor radius is reduced

$$L_{WA} = 50 \cdot \log_{10} V_{tip} + 10 \cdot \log_{10} D - 4\,. \tag{8.2}$$

However, a reduction of rotor frequency and rotor diameter also reduces the power output. This reduction can be estimated for a given turbine if the non-dimensional power curve is known which gives the power coefficient c_P as a function of the tip speed ratio λ.

This estimation has been made for the 2 MW Tjæreborg turbine [94]. Starting from the design tip speed ratio of λ = 7.3, the rotor frequency has been reduced. Figure 8.1 shows that power output changes only slightly, whereas the sound power level decreases linearly with rotor frequency. This behavior is due to the fact that the c_P-λ curve has a flat tangent at the design tip speed ratio. Figure 8.1 shows also that a reduction of rotor diameter reduces the power output drastically and cannot be considered as an appropriate means for noise reduction.

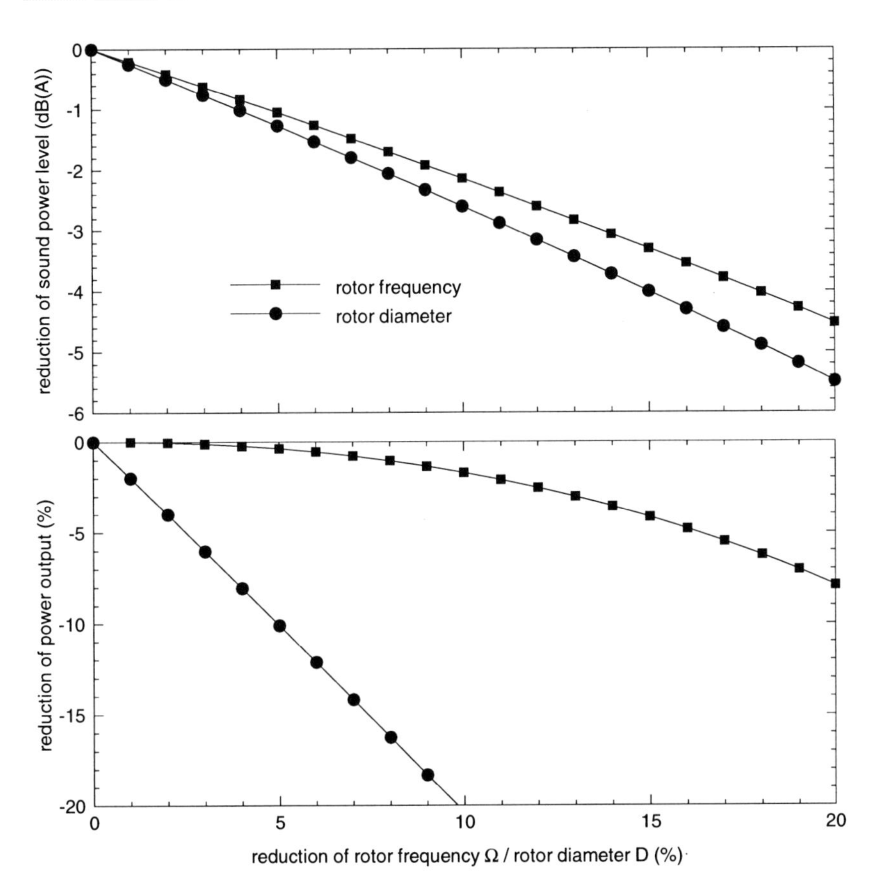

Figure 8.1: Reduction of sound power level and power output with decreasing rotor frequency and rotor diameter.

Another possibility for noise reduction is a modified pitch setting of the blades. From the results of noise measurements on commercial turbines with various pitch settings and from wind tunnel experiments performed within the DEWI project it can be concluded that decreasing the angle of attack by 1° reduces the sound power level by approximately 1 dB(A) [131]. Of course, this will lead again to a reduction of power output.

Figure 8.2 shows calculated power curves of a 600 kW stall-regulated turbine for three different pitch settings. Further simulations show that the total annual energy production is reduced by 1–3 % depending on the average wind speed (see Figure 8.3). However, for stall-regulated turbines pitch variations are limited because of the stall requirements. In most countries, noise regulations foresee lower noise limits during the night. Therefore, it may

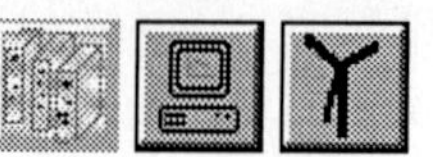

be convenient to take these measures only during night-time in order to minimize the negative impact of reduced rotor frequency and angle of attack on power output [88], [131].

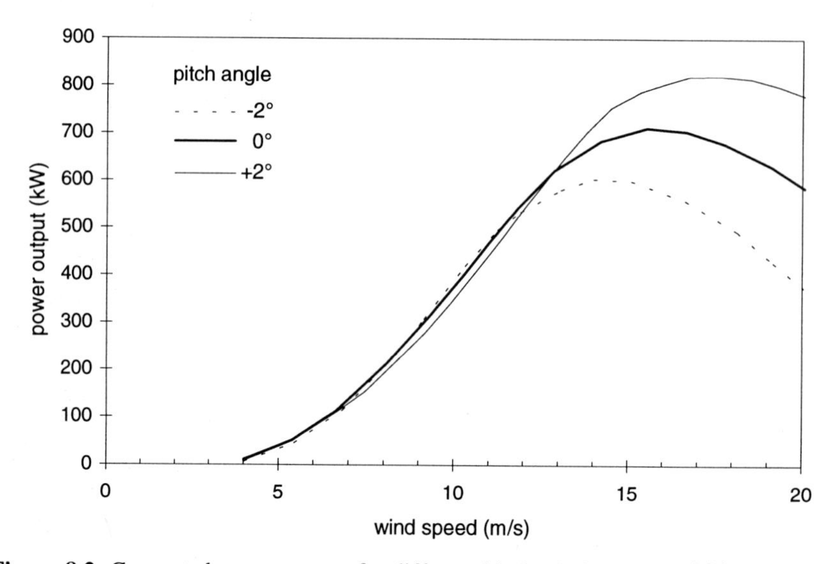

Figure 8.2: Computed power curves for different blade pitch settings [131].

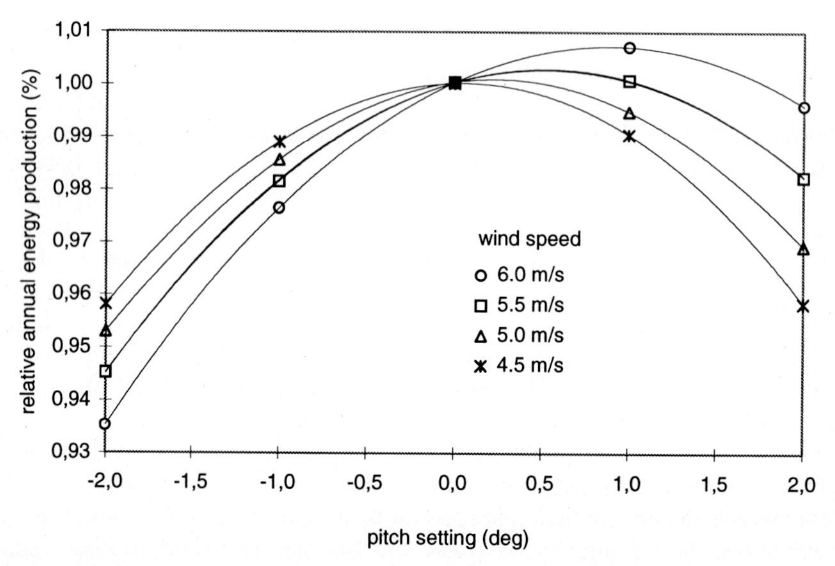

Figure 8.3: Computed annual energy production as a function of blade pitch setting for different average wind speeds at 10 m height [131].

8.3 Reduction of Trailing-Edge Noise

8.3.1 Background

The mechanism of trailing-edge noise and the attempts to model it have been described in Section 3.7. It has been shown that the sound intensity I due to turbulence which is convecting over the edge of an infinitely thin half-plane shows the following dependence

$$I \propto \rho_0 c_0^3 M^5 \alpha^2 \frac{sl}{r^2} \cos^3\left(\overline{\theta}\right) \tag{8.3}$$

where $M = U/c_0$ is the Mach number with the eddy convection velocity U. α is the normalized turbulence intensity, l a length scale of the turbulent region (for example the boundary layer thickness), s the span of the section, and $\overline{\theta}$ the angle between the mean flow direction and the edge (see Figure 3.13). Equation (8.3) shows that the intensity depends on four parameters which can be influenced

- eddy convection velocity U,
- angle $\overline{\theta}$,
- length scale of the turbulent region l,
- normalized turbulence intensity α.

Furthermore, the intensity depends on the radiation efficiency of the edge. The latter can be reduced if the edge is rounded (beveling [109]) or if the acoustic impedance of the edge is reduced (porous, serrated trailing edge [99], [111]).

Trailing edge noise shows a strong dependence on the eddy convection velocity U in the turbulent boundary layer which is reduced compared to the free stream velocity. The most straightforward method of reducing trailing-edge noise is therefore a reduction of the free stream velocity at the blade. This can be accomplished by decreasing the rotor frequency. The noise reduction which can be obtained and the negative effects on the power output are discussed in Section 8.2.

The eddy convection velocity at the trailing edge also depends on the velocity distribution around the airfoil and on the velocity profile in the turbulent boundary layer. Both are influenced by the airfoil shape. However, the airfoil shape influences the value of the normalized turbulence intensity and the boundary layer thickness as well. At present, it is not known which airfoil shape is advantageous for reducing trailing-edge noise.

The strong dependence of intensity on the angle between the mean flow direction and the trailing edge suggests that a reduction might be achieved by giving the blade a swept wing shape. However, this is not practical for wind turbine blades which have a length of 20–30 m.

Table 8.1: Overview of the different means for the reduction of trailing-edge noise.

Modification	*Effect on main parameters*	*Comments*
Reduced rotational speed, blade length	reduces U	see Section 8.2
Swept blade	increases angle α	impractical
Serrated trailing edge	reduces radiation efficiency of trailing edge	see Section 8.3.2
Beveled trailing edge	reduces radiation efficiency of trailing edge	see Sections 8.3.3, 8.4.2
Porous trailing edge	reduces radiation efficiency of trailing edge	see Section 8.3.3
Modified airfoil shape	U, α, and l are influenced	effect on noise is not clear

The radiation efficiency of the edge can be reduced by giving the edge a beveled shape [109] or by reducing the acoustic impedance of the edge. The latter can be accomplished by using porous materials which allow for the acoustic particle velocity to be non-zero at the edge [99]. A second possibility is to give the edge a sawtooth (serrated) shape. Serrated trailing edges have been a major topic of study and are considered as the most promising means for reducing trailing-edge noise [111], [41].

Serrated, beveled, and porous trailing edges have been investigated within the scope of the EU projects (see Preface). These means of noise reduction are described below together with results from experiments. Table 8.1 gives an overview of the different methods for the reduction of trailing-edge noise.

8.3.2 Trailing-Edge Serrations

Two recent wind tunnel investigations [41], [131] have dealt with serrated trailing edge shapes (teeth). Both investigations were initiated by a theoretical description of the noise-reducing mechanism of these edge shapes by Howe [111]. The implications for wind turbine noise reduction were studied in depth by Petterson and reported in [175].

According to Howe, the noise reduction may amount to more than 20 dB, depending on the ratio of the length and the width of the teeth h/λ, the ratio of boundary layer thickness and the tooth length δ/h, and the dimensionless frequency $\omega\delta/U$ based on the boundary layer thickness δ and the eddy convection velocity U at the trailing edge [111]. Figure 8.4 shows the normalized sound pressure spectrum for different ratios h/λ as a function of the dimensionless frequency and a frequency in Hz. For the latter, values of $U = 50$m/s and $\delta = 0.01$ m have been assumed.

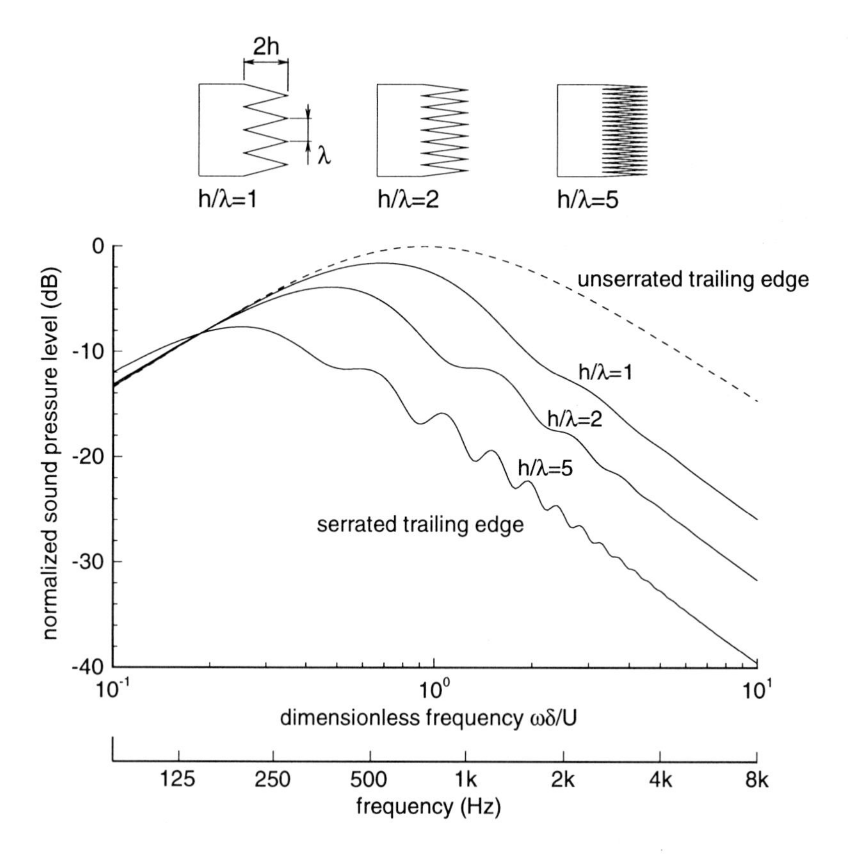

Figure 8.4: Comparison of trailing-edge noise with and without serrations for different ratios of h/λ; $U = 50$ m/s, $\delta = 0.01$ m [111].

Different serrated edge shapes have been investigated within the DEWI project in the wind tunnel of the University of Oldenburg [131] (see Section 7.1). However, the reduction of trailing-edge noise predicted by theory has not been confirmed.

In the framework of the TWIN program, a series of (uncambered) NACA 00xx and (cambered) NACA 63-6xx airfoils with serrated trailing edges were tested in the small anechoic wind tunnel of NLR [41] (see Section 7.1). For both types of airfoil reductions up to 6 dB were established and found to be increasing with tooth length. In contrast to theoretical predictions, the ratio of tooth length to width h/λ was found to be rather insignificant as long as $h/\lambda > 1$. The 6 dB reductions were shown to occur rather independently of flow speed and angle of attack.

If the reduction obtained in a wind tunnel could be realized in free-field operation, an overall reduction of at least 3 dB would result. However, the concept of noise reduction by serrated trailing edges is based on several assumptions with respect to boundary-layer behavior, which are not necessarily valid in the case of a rotating blade. Due to the strong three-dimensional flow around the tip, it may be expected that the specification of the optimal size, spreading and orientation of the teeth at the outer blade section will require a substantial research effort.

In the DELTA project, simple serrated trailing edges were tested on a full-scale turbine, i.e. their size, spreading and orientation was not adjusted to the three-dimensional case. The A-weighted sound power level L_{WA} was normalized to 8 m/s wind speed at 10 m height. L_{WA} was found to be equal to that of a sharp trailing edge, and also the spectra of the noise with the two trailing edges were identical [120].

8.3.3 Modification of Trailing-Edge Shape and Material

Blake [20], [21] investigated the effect of trailing-edge geometry on singing, structural vibration, and broadband noise radiation. He found that an airfoil with a flat pressure side and a beveled suction side gave the most promising results (see Figure 8.5 and Section 4.4). The experiments showed furthermore that flow separation did not occur if the angle ψ was less than 30°.

Howe [109] investigated the effect of beveling on trailing-edge noise. He found that for a given turbulence intensity in the boundary layer, beveling has an influence on noise radiation only at frequencies which are high enough so that the edge can be regarded as a straight-sided wedge over distances of the order of the turbulence length scale. Figure 8.5 shows the normalized sound pressure spectrum for different angles ψ as a function of the dimensionless frequency and a frequency in Hz. For the latter, values of U = 50 m/s and δ = 0.01 m have been assumed. Obviously, a remarkable noise reduction occurs only at very high frequencies.

Although the sound intensity decreases with increasing angle ψ, a practical limit is given by $\psi = 30°$. As stated above, for higher angles flow separation is likely to occur which involves additional turbulence and noise. Furthermore, bluntness noise may occur. Within the ICA project, the effect of trailing-edge beveling has been investigated on the UNIWEX turbine [24]. The results are reported together with those of tip experiments in Section 8.4.2.

Hayden [99] proposed the concept of variable surface impedance for the reduction of trailing-edge noise. The surface impedance is defined as the ratio of the pressure fluctuations on the surface and the normal velocity fluctuations.

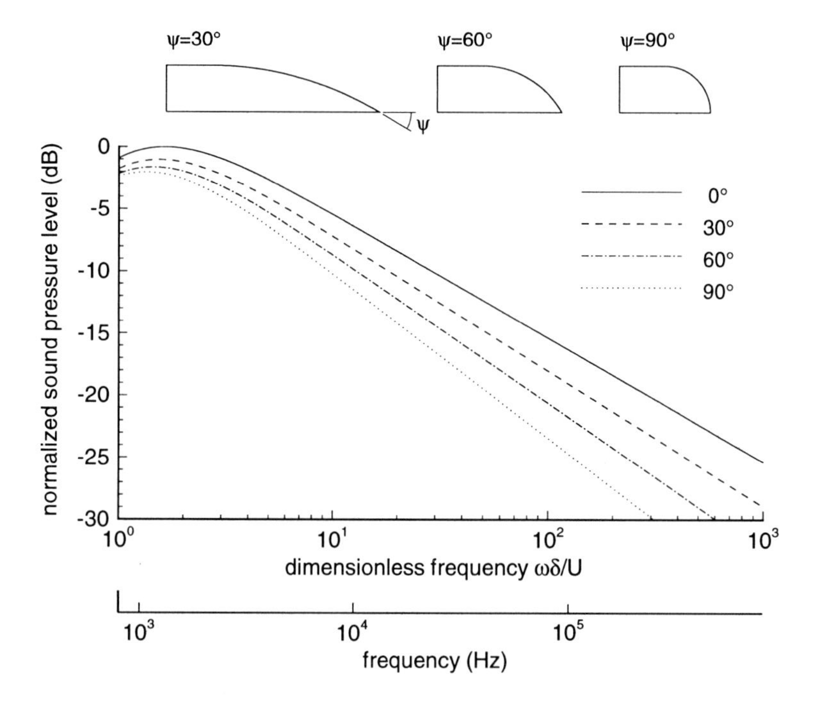

Figure 8.5: Comparison of trailing-edge noise for different beveled trailing edges; $U = 50$ m/s, $\delta = 0.01$ m [109].

Since the latter is zero for rigid surfaces, the surface impedance tends towards infinity. This impedance discontinuity is the reason for the diffraction and amplification of the waves from the turbulent boundary layer. The normal fluid velocity is allowed to be non-zero if the surface is flexible or porous. This measure would reduce the impedance discontinuity and consequently the radiation of sound. Hayden reports experiments with helicopter rotors, powered lift flaps, and simple rotating blades that showed a reduction potential of 5–10 dB [99].

Experiments with a porous trailing edge on a full scale turbine have been performed in the DELTA project. A triangular profile of open cell polyurethane foam was glued to the trailing edge. This increased the chord by 30 mm. The sound power level was found to be 0.5 dB less than that found with a sharp trailing edge. The reduction which is not significant regarding the measurement uncertainty is caused mainly by a decrease of 1–2 dB in the frequency range 500–2000 Hz [120].

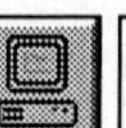

8.4 Reduction of Tip Noise

8.4.1 Background

The first attempts to reduce the aerodynamic noise of wind turbines were very much focused on the shape of the blade 'tips', mainly due to the practical observation that almost all the aerodynamic wind turbine noise emanates from the outer 10 to 20 % of the blade, so that the quieting of this part of the blade would therefore be most effective.

However, even at present, researchers are still in doubt about the extent of extra noise being generated by the three-dimensional flow along the blade in the tip region. Here, it has to be recalled that all known noise mechanisms strongly depend on the effective flow speed of a blade section, resulting in the situation that the tip region will always be the most important contributor, independently of the fact whether three-dimensional flow causes extra noise or not (see Section 3.7.2). Until now outdoor measurement techniques have lacked the resolving power required to answer this question, and theoretical interpretations of the noise-generating mechanism diverge too much so that no definitive conclusions can be drawn.

However, within the scope of the EU projects it has been demonstrated that the change of tip shapes may have a significant influence on the total noise from a wind turbine [24], [120], [131]. Results from acoustic wind-tunnel and outdoor measurements are included in the following.

Within the scope of the ICA project, the literature addressing the tip noise of rotating machinery was investigated [39]. As a result of different Reynolds- and Mach-number ranges, only a small part of the results obtained for fans and helicopters may be applied to wind turbines, leaving only a few reports that give valuable information about tip flow and possible generating mechanisms of tip noise of wind turbines (e.g. [28], [163], [202]).

It was found that tip noise can generally be identified with the turbulence in the locally separated flow region associated with the formation of the cross flow around the tip edge and the formation of the tip vortex. Since the turbulent flow is convected inboard and aft towards the trailing edge, the broadband tip-noise generating mechanism seems to be partially similar to the trailing-edge noise generating mechanism.

Wind tunnel measurements [28] revealed some evidence that for most practical angles of attack the three-dimensional turbulent flow around a (rounded) blade tip increases the levels of trailing-edge noise by some 5 dB. This applies to the higher frequencies (>1000 Hz) where tip noise attained its maximum levels. Over the whole frequency range of interest its contribution is estimated to be limited to 1 to 2 dB(A).

8.4.2 Modification of Tip Shape

ICA project. Based on these insights a blade with an ogee shaped tip planform (ogee tip) and with a shark-fin like tip planform (shark-fin tip) were compared to an elliptical tip, at that time known to be the most silent commercially available tip (reference tip). The shape of both the ogee tip and the shark-fin tip was based on the concept of reaching a lower degree of turbulence in the tip region and reducing the interaction of this turbulence with the trailing edge. See Figure 8.6 and the Annex for a description of the tips.

The tips were mounted as 'gloves' to the blades of the UNIWEX turbine. This experimental 40 kW turbine has a two-bladed rotor of 16 m diameter. The blades have FX79-W151A profiles at the tips. The 'gloves' were adjusted to give the same lifting surface for all three tips. Ogee and reference as well as shark-fin and reference tip were measured simultaneously, using several directional measurement techniques. This procedure allows the tips to be compared during identical flow conditions [24].

The results are shown in Figure 8.6. They reveal that the noise production of the ogee tip is 1–3 dB(A) higher than the noise production of the reference tip. The experiments show furthermore that the shark-fin tip produces more noise than the reference tip. The difference ranges from –1.0 to 5.4 dB(A) and increases with angle of attack. The estimated accuracy of the noise measurement of one blade relative to another is 1 dB.

Measurements with all three tips have been performed in the wind tunnel of the University of Oldenburg (see Chapter 7). Figure 8.6 shows the A-weighted sound pressure levels as a function of angle of attack referred to the results for the reference tip. There is no remarkable difference for the ogee tip. The shark-fin tip is 1–1.5 dB(A) louder for high angles of attack [131].

A second measurement series on the UNIWEX turbine focused on the influence of tip and trailing-edge beveling on noise production. For all three tip planforms mentioned above, several measurements were performed which allowed a direct comparison between a sharp and a beveled trailing edge and between a round and a beveled tip edge. The results of these measurements show that a beveling of a tip edge can lead to a modified sound level spectrum but does not change remarkably the overall sound pressure level.
More promising results were obtained with beveled trailing edges. In Figure 8.7 the differences in sound pressure level between a shark fin tip with a beveled and a sharp trailing edge are shown for different angles of attack. It can be seen that for the frequencies between 2.5 kHz and 5 kHz a reduction of sound pressure level of 3–5 dB is obtained by using a beveled trailing edge.

These results agree qualitatively with those obtained by Howe [109] as described in Section 8.3.3. However, since the main contributions to the overall sound pressure level are generated at frequencies between 500 Hz and 2 kHz, this only results in an entire noise reduction of 1–2 dB.

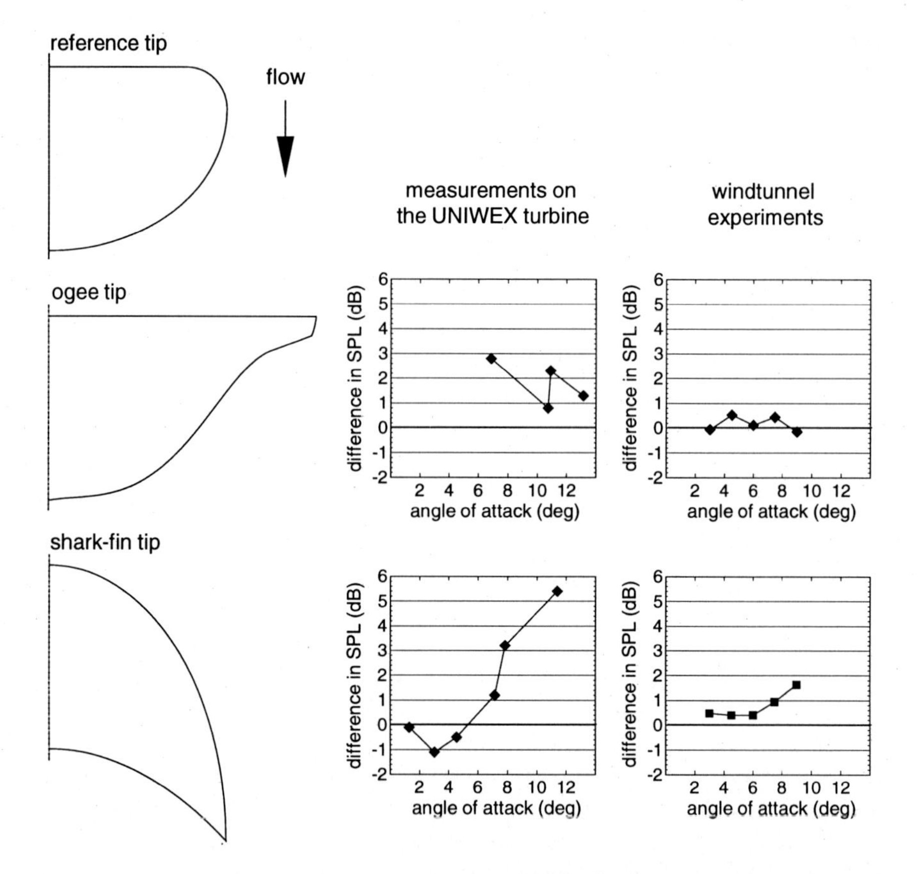

Figure 8.6: Results of tip experiments on the UNIWEX turbine and in the wind tunnel performed within the ICA- and DEWI project [24], [131]; differences in total A-weighted sound pressure level from ogee/shark-fin tip relative to reference tip.

The acoustic measurements for the ogee tip and the reference tip underline this influence of the trailing edge. Further measurements with different rotor diameter and tip velocities have to confirm the results for a wider range of applicability.

DELTA project. In the DELTA project, a series of six tips with different planforms, including an ogee planform, were tested on a Bonus 300 kW turbine [120]. The stall-regulated turbine had an upwind rotor with a diameter of 35 m and was equipped with three LM 17 HHT blades, which used the NACA 63-2xx airfoil series. The tips are shown in Figure 8.8 and described in detail in the Annex. The results were corrected for machinery noise and for the different rotor radii. The measurement uncertainty was estimated to be 1 dB.

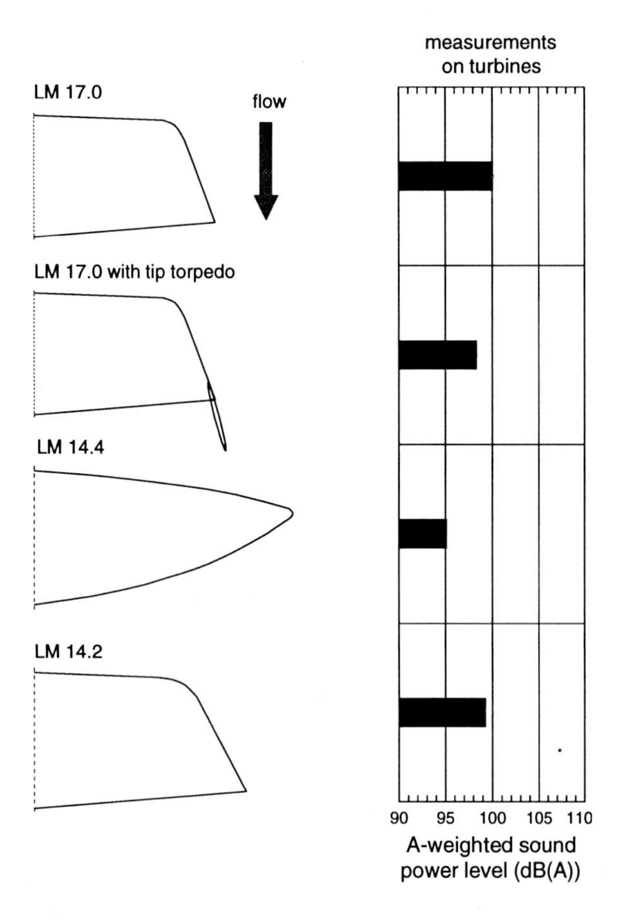

Figure 8.9: Results of tip experiments on commercial turbines performed within the DEWI project [131].

8.5 Reduction of Inflow-Turbulence Noise

Trailing-edge noise is governed by the structure of the boundary-layer turbulence which can be influenced in order to reduce the noise radiation. Atmospheric turbulence cannot be influenced, leaving the response of the blade as the only possibility to reduce inflow-turbulence noise.

As shown by Amiet [3], inflow-turbulence noise can be modeled considering the blades as flat plates if the size of the disturbances is larger than the nose radius of the blade. In this case, the airfoil response function,

which relates the incoming disturbances to the load fluctuations on the blade surface, does not depend on the geometric details of the blade.

At higher frequencies where the size of the disturbances is of the same order as the nose radius, geometric details do affect the noise radiation [126], [127], [128]. However, at present it is not known which airfoil properties are important for a reduction of inflow-turbulence noise.

8.6 Reduction of Blunt-Trailing-Edge Noise

Noise due to blunt trailing edges originates from Karman-like discrete vortex shedding at the trailing edge. It adds a strong tonal component to the spectrum. The noise depends on the thickness of the trailing edge and on the angle of attack. By reducing the thickness the peak frequency is shifted towards higher frequencies and the level is reduced. This has been confirmed in wind-tunnel and outdoor measurements performed within the DEWI project [131]. The results from the wind tunnel measurements are shown in Figure 8.10.

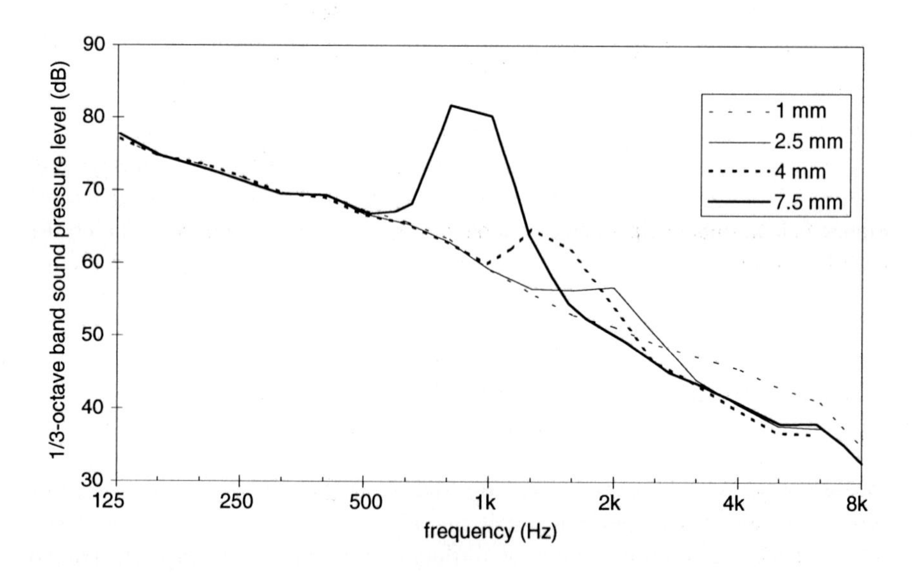

Figure 8.10: Blunt-trailing-edge noise for varying trailing-edge thickness, *Re*=560000 [131].

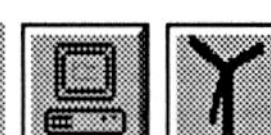

8.7 Conclusions

The most promising concept for the reduction of trailing-edge noise which is the dominating noise mechanism on wind turbines seems to be the serrated trailing edge. Theory predicts a reduction of up to 20 dB depending on the shape of the serrations [111]. A considerable reduction of trailing-edge noise of 6 dB has been found in wind-tunnel experiments performed in the Netherlands [41]. However, in the actual case of a rotating wind turbine blade, the three-dimensional flow at the tip will have an influence on the structure of the boundary-layer turbulence. Up to now, it is not known how serrations have to be orientated in order to take this into account. Furthermore, the problem of production and handling of the blades has to be considered.

First experiments on full scale turbines with non-optimized serrations did not reveal a noise reduction [120]. However, this result is uncertain and further investigations will be needed to clarify this point. So far, no decisive conclusions can be drawn on the potential of other concepts such as beveled or porous trailing edges.

Aeroacoustic theory has given indications that the shape of the tip might have a major influence on the noise radiation. Therefore, several shapes have been investigated within the EU projects. However, it is difficult to draw firm conclusions from the results.

Outdoor experiments with two acoustically optimized tips and a standard elliptical tip showed an increased noise level for the optimized tips [24]. Additional experiments in the wind tunnel revealed no remarkable differences [131]. Outdoor experiments with several tip shapes indicated that an ogee tip (tip O in Figure 8.8) reduced the total sound power level of a turbine by 3–5 dB(A) compared to a standard elliptical tip [120]. Measurements in the wind tunnel showed that the behavior of the tips was completely different at different angles of attack. Here, the ogee tip was better only at high angles [131]. Again, further investigation will be needed. Outdoor experiments with different LM-Glasfiber tips showed encouraging results with a tip shape where both the leading and trailing edge are smoothly cut (see Figure 8.9) [131].

The most straightforward way to reduce aerodynamic noise is a reduction of tip speed and angle of attack. This can be accomplished by reducing the rotor frequency and changing the pitch setting, respectively. A reduction of rotor frequency by 10 % reduces the sound power level by 2–3 dB. The reduction of power output depends on the individual turbine. A reduction of angle of attack by 1° reduces the sound power level by approximately 1 dB and the annual energy production by 1–3 %. In order to minimize the loss of power output it may be convenient to take these measures only at night-time [131].

9 Future Work

This chapter contains a summary of ideas and recommendations for future work in the field of *wind turbine noise*. It is neither complete nor meant as giving the 'final answer' to all questions. However, during reviewing the work performed in JOULE II, it became clear, that there are three important fields for future research on wind turbine noise, namely *modeling*, *reduction*, and *experimental work.*

Modeling. The complete description of wind turbine noise requires the modeling of the *noise generating mechanisms* and its *propagation* towards the observer. In the field of noise prediction improvements are needed for the description of the steady and unsteady flow as well as for the acoustic models. For example, the refined modeling of trailing-edge noise can only be achieved, if the structure of the turbulence, the turbulent kinetic energy, the length scales, the spectral properties, etc. of the flow around an airfoil or rotor blade under given operation conditions can be described in the vicinity of the trailing edge. Being aware of the problem of turbulence modeling, the situation is even more complicated for unsteady flows, which are a typical operational state for a wind turbine rotor and for tip flows. For noise prediction, models that allow the treatment of complete airfoils subjected to turbulent, viscous flow are required instead of using semi-infinite or finite, infinitely thin planes to represent the airfoil [66], [3]. Work in this direction will be done within the JOULE III project DRAW (JOR3-CT95-0083).

In the field of noise propagation the efforts should be directed towards the development of wind turbine siting planning tools for backing the demand for noise legislation which takes into account the differences between 'normal' industrial plants and wind turbines. Advanced propagation models have to be applied for parametric studies on the optimal positioning of wind turbines taking into account wind effects and complex terrain.

Reduction. Theoretical and experimental investigations during previous work [41], [85] have already lead to generation of ideas how to reduce broadband noise by better blade design, i.e.

- serrated trailing edges to reduce trailing-edge noise will be developed towards full-scale commercial application (STENO, JOR3-CT95-0073).
- the influence of blade shape on high-frequency inflow turbulence noise will be explored (DRAW, JOR3-CT95-0083).

The poor understanding of the governing phenomena of tip noise will be elevated to enable optimization of tip plan forms.

Experimental Work. Investigations in a wind tunnel are a basic need for

- the separate study of noise generating mechanisms,
- validation of theories on noise generation and reduction,
- gathering information necessary to develop full-scale reduction concepts by means of detailed flow, performance and acoustic measurements on blade sections in wind tunnels (DRAW, STENO),
- validation and optimization of wind-tunnel-based reduction concepts (STENO)
- validation of propagation models using outdoor flow visualization, performance and acoustic measurements on full-scale turbines.

A lot of the mentioned points are taken from the JOULE III projects DRAW [85] and STENO. The mentioned points can only be detailed to a certain degree, because work on the projects has just started and detailed ideas to reduce noise have to be supported by theoretical work and to be proven by further wind tunnel experiments.

10 References

10.1 Recommended References

This section contains selected references from the list in Section 10.2 that are recommended for further reading. The references cover subjects of almost all chapters of the book and are shortly described below.

1. [22] Blake, W. K.: *Mechanics of Flow-Induced Sound and Vibration, Vol. I: General Concepts and Elementary Sources.* ACADEMIC Press INC., Harcourt Brace Jovanovich, Publishers, pp. 1-425, 1986.

   Blake gives a fairly complete survey of the mechanisms of flow-induced sound and vibration. The two volumes include a development of the important equations of aeroacoustics and a special chapter for the noise mechanisms which are relevant for lifting surfaces in general, namely trailing-edge noise and inflow-turbulence noise.

2. [23] Blake, W. K.: *Mechanics of Flow-Induced Sound and Vibration, Vol. II: Complex Flow-Structure Interactions.* ACADEMIC Press INC., Harcourt Brace Jovanovich, Publishers, pp. 1-973, 1986.

   (description see above)

3. [30] Brooks, F. T.; Pope, D. S.; Marcolini, M. A.: *Airfoil Self-Noise and Prediction.* NASA RP-1218, pp. 1-145, July 1989.

   Brooks et al. report on boundary-layer and acoustic measurements of 2D NACA 0012 airfoil sections and a 3D tip performed in an anechoic wind tunnel at Reynolds number between 400.000–$1 \cdot 10^6$ and angle of attacks between 0–25°. In the analysis they separated trailing-edge noise, stalled flow noise, laminar-boundary-layer-vortex-shedding noise, tip noise, and blunt-trailing-edge noise. A set of spectral scaling formulas was deduced and implemented in a FORTRAN 77 code, which was given in their annex.

4. [75] Gipe, P. B.: *Wind Energy Comes of Age.* John Wiley & Sons, New York, Chichester, Toronto, etc., 1995.

   Gipe chronicles wind energy's progress from its rebirth during the oil crises of the 1970s through a troubling adolescence in California's mountain passes in the 1980s to its maturation on the plains of northern Europe in the 1990s. He cites improvements in the performance, reliability, and cost effectiveness of modern wind turbines to support his contention that wind energy has come of age as a commercial technology for generating electricity.

5. [96] Hayden, R. E.: *Fundamental Aspects of Noise Reduction from Powered-Lift Devices.* SAE Paper 730376, SAE Transactions, pp. 1287-1306, 1973.

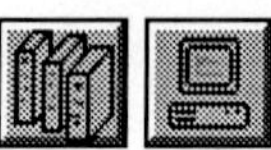

Hayden treats the fundamental aspects of noise reduction from powered lift devices. He gives an overview to the important noise mechanisms (trailing-edge noise, inflow-turbulence noise) and their prediction, and discusses some techniques for noise reduction, for example, the concept of variable impedance.

6. [106] Howe, M. S.: *A Review of the Theory of Trailing Edge Noise*. Journal of Sound and Vibration, Vol. 61, No. 3, pp. 437-465, 1978.

   Howe makes an extensive review of the theory of trailing-edge noise and develops a unified approach which includes all former approaches. He discusses the important features of trailing-edge noise, for example, the question whether a Kutta condition should be applied or not, and derives an expression for the sound radiation in terms of the spectrum of the surface pressure fluctuations.

7. [112] Hubbard, H. H.: *Wind Turbine Acoustics*. NASA Technical Paper 3057, DOE/NASA/20320-77, pp. 1-45, 1990.

   Hubbard and Shepherd give a comprehensive introduction to wind turbine acoustics. They describe the types of noise which are relevant for wind turbines and give noise prediction formulas for the most important mechanisms. Further topics are noise propagation, the noise prediction for multiple turbines, the response of people to noise, and noise measurements.

8. [141] Lighthill, M. J.: *The Bakerian Lecture, Sound Generated Aerodynamically*. Proceedings of the Royal Society of London, Vol. 267, pp. 147-182, 1962.

   Lighthill develops the basic equations for sound generated by turbulence by rearranging the Navier-Stokes equations and showing that the turbulence is equivalent to a volume distribution of acoustic quadrupoles whose strength is proportional to the fluctuating Reynolds-stresses. Lighthill gives a complete overview to his so-called 'acoustic analogy' (see also the original paper [139], [140]).

9. [155] Lowson, M. V.: *Assessment and Prediction of Wind Turbine Noise*. Flow Solutions Report 92/19, ETSU W/13/00284/REP, December 1992.

   Lowson gives a survey of low-frequency and high-frequency noise sources relevant for wind turbines. Three different classes of prediction models are distinguished, representing different levels of complexity. A prediction model including inflow turbulence noise and trailing-edge noise is presented, applied, and results are compared to measured noise spectra.

10. [177] Pinder, J. N.: *Mechanical Noise from Wind Turbines*. Wind Engineering, Vol. 16, No. 3, pp. 158-168, 1992.

    Pinder describes the sources of mechanical noise of HAWT's, their transmission along structures, and their ranking. The major mechanical source, i.e. gearbox noise, is discussed in more detail together with possible means of reduction, both at the source and by modifications of the transmission paths.

10.2 List of References

[1] Ainslie, J. F.; Scott, J.: *Theoretical Modelling of Noise Generated by Wind Turbines*. ECWEC'89, CONF-890717, pp. 458-462, July 1989.

[2] Althaus, D.; Würz, W.: *Windkanaluntersuchungen an drei Blattspitzen für Windturbinen mit horizontaler Achse*. Institut für Aerodynamik und Gasdynamik, November 1994.

[3] Amiet, R. K.: *Acoustic Radiation From an Airfoil in a Turbulent Stream*. Journal of Sound

and Vibration. Vol. 41, No. 4, pp. 407-420, April 1975.

[4] Amiet, R. K.: *Noise Due To Turbulent Flow Past a Trailing Edge*. Journal of Sound and Vibration, Vol. 47, No. 3, pp. 387-393, March 1976.

[5] Amiet, R. K.: *Effect of the Incident Surface Pressure Field on Noise Due to Turbulent Flow Past a Trailing Edge*. Journal of Sound and Vibration, Vol. 57, No. 2, pp. 305-306, 1978.

[6] Andersen, B.; Jakobsen, J.: *Noise from Wind Turbines. Noise from Wind Farms and Propagation of Noise from Wind Turbines (in Danish)*. Report 70.89.783.1/ LI 510/92, Danish Acoustical Institute, Lyngby 1992.

[7] Anonym: *Guide for measuring and calculating industrial noise (in Dutch)*. Ministerie van Volksgezondheit en Milieuhygiëne, Interdepartementale Commissie Geluidhinder (ICG), IL-HR-13-01, 1981.

[8] Anonym: *Gesetz über die Einspeisung von Strom aus erneuerbaren Energien in das öffentliche Netz (Stromeinspeisegesetz, Energy Feeding Law, EFL)*. Federal law Germany, forced to action 7 December 1990.

[9] Anonym: *Publication of Research and Technological Development Programme in the Field of Non-Nuclear Energy; JOULE II*. EUR 15893 EN, 1994.

[10] Ardoullie, M.: *Wind Energy on Small Grids*. European Directory of Renewable Energy, James & James Science Publishers Ltd., ISBN 1-873936-16-8, ISSN 0266-8041, pp. 199-200, 1993.

[11] Banaugh, R. P.; Goldsmith, W.: *Diffraction of Steady Acoustic Waves by Surfaces of Arbitrary Shape*. Journal of the Acoustical Society of America, Vol. 35, No. 10, pp. 1590-1601, October 1963.

[12] Bareiß, R.; Wagner, S.: *Load Calculations on Rotor Blades of a Wind Turbine*. Proc. of the 5th IEA Symposium, University of Stuttgart, pp. 1-10, December 1991.

[13] Bareiß, R.; Wagner, S.: *A Hybrid Wake Model for Wind Turbines*. STAB Symposium Köln-Porz, DGLR-Report 92-07, 1992.

[14] Bareiß, R.; Wagner, S.: *The Free Wake/Hybrid Wake Code ROVLM - A Tool for Aerodynamic Analysis of Wind Turbines*. Stephans, H.S. (ed.), Proc. of the European Community Wind Turbine Conference, pp. 424-431, March 1993.

[15] Bareiß, R.; Wagner, S.: *Aerodynamic Codes as a Basis for Aerodynamic Noise Prediction of Horizontal Axis Wind Turbines*. Proceedings NOISE-93, St. Peterburg, Crocker, M. J.; Ivanov, N. I. (ed.), Vol. 1, pp. 31-36, May - June 1993.

[16] Bareiß, R.; Guidati, G.; Wagner, S.: *An Approach Towards Refined Noise Prediction of Wind Turbines*. Proc. of the European Wind Energy Association Conf.& Exhibiton, Dr. J. L. Tsipouridis (ed.), Thessaloniki, Vol. 1, pp. 785-790, October 1994.

[17] Barman, K.; Dahlberg, J. A.; Meijer, S.: *Measurement of the Tower Wake of the Swedish Prototype WECS Maglarp and Calculations of its Effect on Noise and Blade Loading*. Proceedings of the European Wind Energy Conference, pp. 56-63, October 1984.

[18] Beiser, A.: *Physics*. The Benjamin/Cummings Publishing Company Inc., Menlo Park, California, USA, 1982.

[19] Best, R.: *Digitale Meßwertverarbeitung*. R. Oldenburg Verlag München, Wien, 1991.

[20] Blake, W. K.: *Excitation of plates and hydrofoils by trailing edge flows*. Turbulence-induced Vibrations and Noise of Structures, New York: ASME, 1983.

[21] Blake, W. K.: *Aero-hydroacoustics for ships*. David Taylor Naval Ship Research and Development Center, Report No. 84/010, 1984.

[22] Blake, W. K.: *Mechanics of Flow-Induced Sound and Vibration, Vol. I: General Concepts and Elementary Sources*. ACADEMIC Press INC., Harcourt Brace Jovanovich, Publishers, pp. 1-425, 1986.

[23] Blake, W. K.: *Mechanics of Flow-Induced Sound and Vibration, Vol. II: Complex Flow-Structure Interactions*. ACADEMIC Press INC., Harcourt Brace Jovanovich, Publishers,

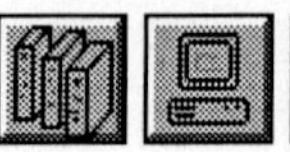

pp. 426-973, 1986.

[24] Braun, K. et al.: *Investigation of blade tip modifications for acoustic noise reduction and rotor performance improvement.* Final report, JOUR-CT90-0111 and JOU2-CT92-0205, Institut für Computer Anwendungen (ICA), Universität Stuttgart, 1995. to be published.

[25] Brooks, T. F.; Hodgson, T. H.: *Prediction and Comparison of Trailing Edge Noise Using Measured Surface Pressures.* AIAA 6th Aeronautics Conference, AIAA Paper 80-0977, pp. 1-30, June 1980.

[26] Brooks, T. F.; Hodgson, T. H.: *Trailing Edge Noise Prediction from Measured Surface Pressures.* Journal of Sound and Vibration, Vol. 78, No. 1, pp. 69-117, March 1981.

[27] Brooks, T. F.; Marcolini, M. A.: *Scaling of Airfoil Self-Noise Using Measured Flow Parameters.* AIAA Journal, Vol. 23, No. 2, pp. 207-213, February 1985.

[28] Brooks, T. F.; Marcolini, M. A.: *Airfoil Tip Vortex Formation Noise.* AIAA Journal, Vol. 24, No. 2, pp. 246-252, February 1986.

[29] Brooks, T. F.; Marcolini, M. A.; Pope, D. S.: *Airfoil Trailing-Edge Flow Measurements.* AIAA Journal, Vol. 24, No. 8, pp. 1245-1251, August 1986.

[30] Brooks, F. T.; Pope, D. S.; Marcolini, M. A..: *Airfoil Self-Noise and Prediction.* NASA RP-1218, pp. 1-137, July 1989.

[31] Brown, N. A.: *Noise and Fluids Engineering.* ASME, Winter Annual Meeting, Atlanta, GA.

[32] Bruggeman, J. C.; Parchen, R. R.: *Reduction of Broadband Noise of Wind Turbines and Computational Fluid Dynamics.* Proc. 2nd European Computational Fluid Dynamics Conference, ECCOMAS, Stuttgart, John Wiley and Sons, pp 145-148, 1994.

[33] Chapman, J.: *The USA Experience.* Proc. of the 5th European Wind Energy Association Conference, Thessalonoki, Tsipouridis, J. L. (ed.), Vol. I, pp. 33-36, October 1994.

[34] Ciskowsky, R. D.; Brebbia, C. A.: *Boundary Element Methods in Acoustics.* Computational Mechanics Publications, Elsevier Applied Science, 1991.

[35] Concawe Report 4/81: *The Propagation of Noise from Petrochemical Complexes to Neighbouring Communities.* Den Haag, 1981.

[36] Corbet, D.: *Wind Farms in Areas of Low Wind Speed.* European Directory of Renewable Energy, James & James Science Publishers Ltd, ISBN 1-873936-16-8, ISSN 0266-8041, pp. 188-194, 1993.

[37] Curle, N.: *The Influence of Solid Boundaries Upon Aerodynamic Sound.* Proceedings of the Royal Society, Ser. A, Vol. 231, pp. 506-514, 1955.

[38] Danish Acoustical Laboratory: *Environmental noise from industrial plants - General prediction method.* Report 32, Lyngby, 1981.

[39] Dassen, A. G. M.: *Tip Noise of Wind Turbines, Results of a Literature Study and Proposals for 'Silent' Blade Tips.* NLR contract report CR 92407 L, 1992.

[40] Dassen, A. G. M.: *Experimental Validation of a Wind Turbine Aerodynamic Noise Prediction Code.* ECWEC'93, Proc. of the European Community Wind Energy Conference, Germany, Stephens, H. S. (ed.), March, 1993.

[41] Dassen, A. G. M.; Parchen, R.; Bruggeman, J.; Hagg, F.: *Wind Tunnel Measurements of the Aerodynamic Noise of Blade Sections.* Proc. of the European Wind Energy Association Conf.& Exhibiton, Dr. J. L. Tsipouridis (ed.), Thessaloniki, Vol. I, pp. 791-798, October 1994.

[42] Dassen, A. G. M.: *Personal Communication, 1996.*

[43] Delaney, M. E.; Bazley, E. N.: *Acoustical Properties of Fibrous Absorbent Materials.* Applied Acoustics, Vol. 3, pp. 105-116, 1970.

[44] De Wolf, W. B.: *EEN Predictiemethode Voor Het Aerodynamische Geluid Van Windturbines Met Horizontale AS.* NLR TR 87018 L, pp. 1-55, December 1986.

[45] De Wolf, W. B.: *Aerodynamisch Geluid van Windturbines*. NLR MP 87004 U, pp. 1-22, February 1987.

[46] Dexin, H.; Ming, C.; Dahlberg, J. A.; Ronsten, G.: *Flow Visualization on a Rotating Wind Turbine Blade*. FFA TN 1993-28, 1993.

[47] Diamantaras, K.; Palz, W.: *The European Community Wind Energy R & D Programme*. EWEC '94, Proc. 5th European Wind Energy Association Conference and Exhibition, Thessaloniki, Tsipouridis J. L.(ed.), Vol. I, pp. 105-110, October, 1994.

[48] Drela, M.: *XFOIL: An Analysis and Design System for Low Reynolds Number Airfoils*. Conference on Low Reynolds Number Aerodynamics, University Notre Dame, June 1989.

[49] Dunbabin, P.: *Aerodynamic Noise Emission from Wind Turbines; Scrutiny of Experimental Data*. pp. 1-10, March 1993.

[50] Dunbabin, P.: *Aerodynamic Noise Prediction from Wind Turbines*. March 1993.

[51] Edwards, P. D.: *The Performance and Problems of and the Public Attitude to the Delabole Wind Farm*. In Pitcher, K. F. (ed.) Wind Energy Conversion, pp. 23-25, 1993.

[52] ESDU Data Sheets: *Characteristics of Atmospheric Turbulence Near the Ground Part I: Definitions and General Information*. Engineering Sciences Data Item No. 74030, pp. 1-16, October 1976.

[53] ESDU Data Sheets: *Characteristics of Atmospheric Turbulence Near the Ground Part II: Single Point Data for Strong Winds*. Engineering Sciences Data Item No. 85020, pp. 1-37, November 1990.

[54] ESDU Data Sheets: *Characteristics of Atmospheric Turbulence Near the Ground; Part III: Variations in Space and Time for Strong Winds*. Engineering Sciences Data Item No. 86010, pp. 1-34, September 1991.

[55] Farassat, F.: *Theory of Noise Generation from Moving Bodies With an Application to Helicopter Rotors*. NASA TR R-451, pp. 1-59, December 1975.

[56] Farassat, F.: *Discontinuities in Aerodynamics and Aeroacoustics: The Concept and Applications of Generalized Derivatives*. Journal of Sound and Vibration, Vol. 55, No. 2, pp. 165-193, 1977.

[57] Farassat, F.; Succi, G. P.: *A Review of Propeller Discrete Frequency Noise Prediction Technology with Emphasis on two Current Methods for Time Domain Calculations*. Journal of Sound and Vibration, Vol. 71, No. 3, pp. 399-419, 1980.

[58] Farassat, F.: *Linear Acoustic Formulas for Calculation of Rotating Blade Noise*. AIAA Journal, AIAA80-0996R, Vol. 19, No. 9, pp. 1122-1130, September 1981.

[59] Farassat, F.; Myers, M. K.: *Extension of Kirchhoff's Formula to Radiation from Moving Surfaces*. Journal of Sound and Vibration, Vol. 123, No. 3, pp. 451-461, 1988.

[60] Farassat, F.: *Introduction to Generalized Functions With Applications in Aerodynamics and Aeroacoustics*. NASA TP 3428, WU 535-03-11-02, pp. 1-45, May 1994.

[61] Fileds, J. M.: *A Quantitative Summary of Non-Acoustical Variables' Effect on Reactions to Environmental Noise*. Noise Con 90. University of Texas, Austin, pp. 303-308, 1990.

[62] Fink, M. R.: *Experimental Evaluation of Theories for Trailing Edge and Incidence Fluctuation Noise*. AIAA Journal, Vol. 13, No. 11, pp. 1472-1477, November 1975.

[63] Fink, M. R.; Bailey, D. A.: *Airframe Noise Reduction Studies and Clean-Airframe Noise Investigation*. NASA CR-159311, 1980.

[64] Ffowcs-Williams, J. E.: *Thoughts on the Problem of Aerodynamic Noise Sources near Solid Boundaries*. AGARD Report 459, April 1963.

[65] Ffowcs Williams, J. E.; Hawkins, D. L.: *Sound Generation by Turbulence and Surfaces in Arbitrary Motion*. Philosophical Transactions of the Royal Society of London, Vol. 264, No. A 1151, pp. 321-42, May 1969.

[66] Ffowcs Williams, J. E.; Hall, L. H.: *Aerodynamic Sound Generation by Turbulent Flow in the Vicinity of a Scattered Half Plane*. Journal of Fluid Mechanics, Vol. 40, No. 4, pp. 657-

670, 1970.

[67] Ffowcs Williams, J. E.: *Aeroacoustics*. Annual Review of Fluid Mechanics, No. 9, pp. 447-468, 1977.

[68] Ffowcs Williams, J. E.: *Noise Source Mechanisms in Unsteady Flow*. Acustica, Vol. 50, No. 3, pp. 167-179, 1982.

[69] Garrad, A. D.: *Wind energy in Europe - Time for action*, EWEA Strategy document, October 1991, Page 1-59.

[70] Garrad, A.: *Where We are Now*. EWEC '94, Proceedings of 5th European Wind Energy Association Conference and Exhibition, Thessaloniki, Tsipouridis J. L. (ed.), Vol. I, pp. 21-32, October, 1994.

[71] Gasch, R.: *Windkraftanlagen*. B. G. Teubner Verlag, Stuttgart, pp. 1-371, 1993.

[72] Gaudiosi, G.; Cesari, F.: *Offshore Wind Energy Resource in the Mediterranean*. EWEC '94, Proc. 5th European Wind Energy Association Conference and Exhibition, Thessaloniki, Tsipouridis J. L.(ed.), Vol. I, pp. 116-122, October 1994.

[73] Gipe, P. B.: *Overview of Worldwide Wind Generation*. World Wide Web, http://keynes.fb12.tu-berlin.de/Luftraum/Konst/pgipe.html, 1994.

[74] Gipe, P. B.: *Design as if People Matter: Aesthetic Guidelines for the Wind Industry*. World Wide Web, http://keynes.fb12.tu-berlin.de/Luftraum/Konst/pgipe.html, 1995.

[75] Gipe, P. B.: *Wind Energy Comes of Age*. John Wiley & Sons, New York, Chichester, Toronto, 1995.

[76] Glegg, S. A. L.: *Significance of Unsteady Thickness Noise Sources*. AIAA Journal, Vol. 25, No. 6, pp. 839-844, June 1987.

[77] Glegg, S. A. L.; Baxter, S. M.; Glendinning, A. G.: *The Prediction of Broadband Noise from Wind Turbines*. Journal of Sound and Vibration, Vol. 118, No. 2, pp. 217-239, 1987.

[78] Goldstein, M. E.: *Aeroacoustics*. Mc-Graw-Hill, New York, 1976.

[79] Green, C. D.: *Integral Equation Methods*. Thomas Nelson and Sons Ltd., London, Great Britain, 1969.

[80] Greene, G. C.; Hubbard, H. H.: *Some Calculated Effects of Non-Uniform Inflow on the Radiated Noise of a Large Wind Turbine*. TM-81813, pp. 1-14, May 1980.

[81] Greene, G. C.: *Measured and Calculated Characteristics of Wind Turbine Noise*. NASA CP2185, pp. 355-362, February 1981.

[82] Grosveld, F. W.: *Prediction of Broadband Noise from Horizontal Axis Wind Turbines*. Journal of Propulsion and Power. Vol. 1, No. 4, pp. 292-299, July 1985.

[83] Guidati, G.; Bareiß, R.; Wagner, S.: *Steps Towards Refined Noise Prediction of Wind Turbines*. Proc. of the 7th IEA Symposium "Joint Action Aerodynamics", Lyngby, November 1993.

[84] Guidati, G.; Bareiß, R.; Wagner, S.: *An Investigation of Blade-Tower-Interaction Noise (BTI) for Horizontal Axis Wind Turbines in Upwind and Downwind Configuration First Step Towards Modeling of Aeroelastic Effects*. Proc. of the 8th IEA Symposium, "Joint Action, Aerodynamics of Wind Turbines", Lyngby, November 1994.

[85] Guidati, G.; Bareiß, R.; Wagner, S. et al.: *Development of Design Tools for Reduced Aerodynamic Noise Wind Turbines (DRAW)*. to be published on the European Union Wind Energy Conference and Exhibition, Göteborg, 20-24 May, 1996.

[86] Gutenberg, B.: *Propagation of Sound Waves in the Atmosphere*. The Journal of the Acoustical Society of America, Vol. 13, pp. 151-155, October 1942.

[87] Gutin, L.: *On the Sound Field of a Rotating Propeller*. NACA TM 1195, 1948.

[88] Hagg, F.: *Aerodynamic Noise Reduced Design of Large Advanced Wind Turbines*. ECWEC'90, Proc. of the European Community Wind Energy Conference, Madrid, Spain, pp. 384-388, September 1990.

[89] Hagg, F.; van der Borg, N. J. C. M.; Bruggeman, J. C.; et al.: *Definite Aero-Geluidonderzoek Twin*. Stork Product Engineering B.V., SPE 92-025, April 1992.

[90] Hagg, F.; Bruggeman, J. C.; Dassen, A. G. M.: *National Aero-acoustic Research on Wind Turbines in the Netherlands*. ECWEC'93, Proceedings of the European Community Wind Energy Conference and Exhibition, pp. 290-293, March 1993.

[91] Hansen, A. C.: *Aerodynamics of Horizontal-Axis Wind Turbines*. Annual Review of Fuild Mechanics, Vol. 25, pp. 115-149, 1993.

[92] Hau, E.: *Windkraftanlagen; Grundlagen, Technik, Einsatz, Wirtschaft-lichkeit*. Springer-Verlag, New York, pp. 1-683, 1988.

[93] Hau, E.; Snel, H.; Foellings, F.; et al.: *Study on the Next Generation of Large Wind Turbines*. ECWEC'90, Proc. European Community Wind Energy Conference, Spain, pp. 427-447, September 1990.

[94] Hau, E., Langenbrinck, J., Palz, W.: *WEGA Large Wind Turbines*. Springer-Verlag, Berlin, pp. 1-143, 1993.

[95] Hawkins, J. A.: *Application of Ray Theory to Propagation of Low Frequency Noise from Wind Turbines*. NASA-CR-178367, N88-12349, pp. 1-95, July 1987.

[96] Hayden, R. E.: *Fundamental Aspects of Noise Reduction from Powered-Lift Devices*. SAE Paper 730376, SAE Transactions, pp. 1287-1306, 1973.

[97] Hayden, R. E.; Chanaud, R. C.: *Methods of Reducing Sound Generation in Fluid Flow Systems Embodying Foil Structures and the Like*. United States Patent 3.779.338, December 1973.

[98] Hayden, R. E.; Chanaud, R. C.: *Foil Structures with Reduced Sound Generation*. United States Patent 3.853.428, December 1974.

[99] Hayden, R. E.; Aravamudan, K. S.: *Prediction and Reduction of Rotor Broadband Noise*. No. 1, pp. 61-87, May 1978.

[100] Haykin, S.: *Array Signal Processing*. Englewood Cliffs, N. J. Prentice Hall, 1985.

[101] Hersh, A. S.; Hayden, R. E.: *Aerodynamic Sound Radiation from Lifting Surfaces with and without Leading-Edge Serration*, NASA-CR-114370, 1971.

[102] Hersh, A. S.; Soderman, P. T.; Hayden, R. E.: *Investigation of Acoustic Effects of Leading Edge Serrations on Airfoils*. Journal of Aircraft, Vol. 11, No. 4, pp. 197-201, April 1974.

[103] Hinsch, C.; Schetelich, R.: *On the way to a "Sustainable Development"?* DEWI Magazin, ISSN 0946-1787, Vol. 4, No. 7, pp. 28-34, August 1995.

[104] Hinze, J. O.: *Turbulence*. 2nd ed., McGraw-Hill, 1975.

[105] Howe, M. S.: *Contributions to the Theory of Aerodynamic Sound, with Applications to Excess Jet Noise and the Theory of the Flute*. Journal of Fluid Mechanics, Vol. 71, No. 4, pp. 625-673, 1975.

[106] Howe, M. S.: *A Review of the Theory of Trailing Edge Noise*. Journal of Sound and Vibration, Vol. 61, No. 3, pp. 437-465, 1978.

[107] Howe, M. S.: *Aerodynamic Sound Generated by a Slotted Trailing Edge*. Royal Society Proceedings Series A-Mathematical and Physical Sciences, Vol. 373, No. 1753, pp. 235-252, November 1980.

[108] Howe, M. S.: *On the Generation of Side-Edge Flap Noise*. Journal of Sound and Vibration, Vol. 80, No. 4, pp. 555-573, July 1982.

[109] Howe, M. S.: *Contributions to the Theory of Sound Production by Vortex-Airfoil Interaction, with Application to Vortices with Finite Axial Velocity Defect*. Proceedings of Royal Society London, Vol. A 420, pp. 157-182, January 1988.

[110] Howe, M. S.: *The Influence of Surface Rounding on Trailing Edge Noise*. Journal of Sound and Vibration, Vol. 126, No. 3, pp. 503-523, 1988.

[111] Howe, M. S.: *Noise Produced by a Sawtooth Trailing Edge*. The Journal of the Acoustical

Society of America, Vol. 90, No. 1, pp. 482-487, July 1991.

[112] Hubbard, H. H.; Shepherd, K.P.: *Wind Turbine Acoustics.* NASA Technical Paper 3057, DOE/NASA/20320-77, pp. 1-45, 1990.

[113] Hubbard, H. H.; Shepherd, K. P.: *Aeroacoustics of Large Wind Turbines.* The Journal of the Acoustical Society of America, Vol. 89, No. 6, pp. 2495-2507, June 1991.

[114] IEC-Standard (Draft): *Acoustic Noise Measurement Techniques for Wind Turbine Generator Systems.* 8th working draft, TC 88 WG5(Dubois)16, March 1994.

[115] ISO 266: *Akustik - Normfrequenzen.* pp. 1-4, November 1994.

[116] ISO 9613-1: *Acoustics - Attenuation of Sound during Propagation Outdoors - Part 1. Calculation of the Absorption of Sound by the Atmosphere.* 1993.

[117] ISO 9613-2 (Draft): *Acoustics - Attenuation of Sound during Propagation Outdoors - Part 2. General method of calculation.* 1995.

[118] Israelsson, S.: *Sound Propagation in the Atmosphere near the Ground* (in Swedish), University of Uppsala, 1979.

[119] Jakobsen J.: *Noise from Wind Turbine Generators. Noise Control, Propagation, and Assessment.* Proceeding of the Inter-Noise 90 Conference, pp. 303-308, 1990.

[120] Jakobsen J.; Andersen B.: *Aerodynamical Noise from Wind Turbine Generators; Experiments with Modification of Full Scale Rotors.* Danish Acoustical Institute, EFP j.nr. 1364/89-5 JOUR-CT 90-0107, pp. 1-97, June 1993.

[121] Johnson, W.: *Helicopter Theory.* ISBN 0-691-07917-4, Princeton University Press, New Jersey, 1980.

[122] Information Package: *R & D Activities in the Field of Non-Nuclear Energy (JOULE-THERMIE) 1994-1998.* European Commission, DG XII, DG XVII, Edition 1994-1995.

[123] Kehrbaum, R.: *Perspectives for the Recycling of Wind Turbines.* DEWI Magazin, ISSN 0946-1787, Vol. 4, No. 7, pp. 35-38, August 1995.

[124] Kelley, N. D.; McKenna, H. E.: *Acoustic Noise Associated with the Mod-1 Wind Turbine: Its Source, Impact, and Control.* SERI/TR-635-1166, pp. 1-262, February, 1985.

[125] Kendall, J. M.; Ahtye, W. F.: *Noise Generation by a Lifting Flap/Wing Combination at Reynolds Numbers up to $2.8 \cdot 10^6$.* AIAA Paper 80-0035, 1980.

[126] Kerschen, E. J.; Myers, M. R.: *Incidence Angle Effects on Convected Gust Airfoil Noise.* 8th Aeroacustic Conference, Atlanta, GA, April 11-13, 1983, AIAA Paper 83-0765, pp. 1-11, May 1983.

[127] Kerschen, E. J.; Myers, M. R.: *Influence of Airfoil Mean Loading on Convected Gust Interaction Noise.* Proceedings of the Symposium on Aero- and Hydro-acoustics, France, Springer-Verlag, Berlin and New York, pp. 13-20, July 1985.

[128] Kerschen, E. J.; Myers, M. R.: *A Parametric Study of Mean Loading Effects on Airfoil Gust Interaction Noise.* AIAA, 11th Aeroacoustics Conference, CA, Oct. 19-21, AIAA Paper 87-2677, pp. 11, October 1987.

[129] Klug, H.: *Sound-Speed Profiles Determined from Outdoor Sound Propagation Measurements.* The Journal of the Acoustical Society of America, Vol. 90, No. 1, pp. 475-481, July 1991.

[130] Klug, H.; Gabriel, J.; Osten, T.: *Sound Power Level Measurements in Combination With Electric Power Output Measurements.* Proc. of the European Wind Energy Association Conf.& Exhibiton, Dr. J. L. Tsipouridis (ed.), Thessaloniki, Vol. I, pp. 165-168, October 1994.

[131] Klug, H.; Osten, T.; Jakobsen, J.; Andersen, B.; et al.: *Aerodynamic Noise from Wind Turbines and Rotor Blade Modification.* JOULE II, Project JOU2-CT92-0233, Final Report, DEWI-V-950006, November 1995.

[132] Kragh, J.: *Noise from Industrial Plants; Measurements and Prediction.* Nordfosk, Stockholm, pp. 1-110, 1984.

[133] Kragh, J.: *Propagation of Wind Turbine Noise. A Preliminary Note*. Technical Report, AV 427/94, 1994.

[134] Kragh, J.; Jakobsen, J.: *Propagation of Wind Turbine Noise, Outline of a Prediction Method*. DELTA Acoustics & Vibration, AV 320/95, pp. 1-30, 1995.

[135] Lahti, T.: *Calculation of Noise Propagation with Multiple Ground Reflections in an Inhomogeneous Atmosphere*. Proc. Internoise 81, Amsterdam 1981.

[136] Lamb H.: *Hydrodynamics*. pp. 1-738, Dover Publications, New York, April 1932.

[137] Larsson, C.; Israelsson, S.: *The Effects of Meteorological Parameters on Sound Propagating from a Point Source*. Report No. 67, 1982.

[138] Lee, T.; Wren, B.; Hickman, M.: *Public Responses to the Siting and Operation of Wind Turbines*. Robens Institute and Department of Psychology, University of Surry, 1989.

[139] Lighthill, M. J.: *On Sound Generated Aerodynamically; I. General Theory*. Proceedings of the Royal Society of London. Series A, Vol. 211, pp. 564-587, 1952.

[140] Lighthill, M. J.: *On Sound Generated Aerodynamically, II Turbulence as a Source of Sound*. Proceedings of the Royal Society of London. Series A, Vol. 222, pp. 1-32, 1954.

[141] Lighthill, M. J.: *The Bakerian Lecture, Sound Generated Aerodynamically*. Proceedings of the Royal Society of London. Series A, Vol. 267, pp. 147-182, 1962.

[142] Lighthill, M. J.: *Waves in Fluids*. Cambridge University Press, Cambridge, London, New York, Melbourne, ISBN 0-521-21689-3, 1978.

[143] Lighthill, M. J.: *Jet Noise*. AGARD Report 448, April 1963.

[144] Lighthill, M. J.: *Introduction to Fourier Analysis and Generalised Functions*. Cambridge University Press, London, New York, 1964.

[145] Ljunggren, S.: *Recommendation Practices for Wind Turbine Testing, 4. Acoustics. Measurements of Noise Emission From Wind Turbines*. 3. Edition 1994, The Royal Inst. of Technology, Sweden, pp. 1-54, 1994.

[146] Lohmann, D.: *Prediction of Ducted Radiator Fan Aerocoustics*. DGLR/AIAA 14th Aeroacoustics Conference, Aachen, DGLR-Bericht 92-02, DLR/AIAA 92-02-098, 1992.

[147] Lölgen, Th.; Neuwerth, G.: *Noise Emission of Propfans Due to Inflow Distortions*. DGLR/AIAA 92-02-050, 14th Aeroacoustics Conference, pp. 325-333, 1992.

[148] Lowson, M. V.: *The Sound Field for Singularities in Motion*. Proc. Royal Society (London), Series A, Vol. 286, pp. 559-572, 1965.

[149] Lowson, M. V.: *Basic Mechanisms of Noise Generation by Helicopters, V/STOL Aircraft, and Ground Effect Machines*. Wyle Laboratories -Research Staff, Report WR 65-9, pp. 1-32, May 1965.

[150] Lowson, M. V.: *Reduction of Compressor Noise Radiation*. The Journal of the Acoustical Society of America, Vol. 43, No. 1, pp. 37-50, 1968.

[151] Lowson, M. V.: *Theoretical Studies of Compressor Noise*. NASA CR-1287, pp. 1-100, March 1969.

[152] Lowson, M. V.: *Theoretical Analysis of Compressor Noise*. The Journal of Acoustical Society of America, Vol. 47, No. 1, Part 2, pp. 371-385, 1970.

[153] Lowson, M. V.: Whatmore, A. R.; Whitfield, C. E.: *Source Mechanisms for Rotor Noise Radiation*. NASA CR 2077, 1973.

[154] Lowson, M. V.: *Applications of Aero-Acoustic Analysis to Wind Turbine Noise Control*. Wind Engineering, Vol. 16, No. 3, pp. 126-140, 1992.

[155] Lowson, M. V.: *Assessment and Prediction of Wind Turbine Noise*. Flow Solutions Report 92/19, ETSU W/13/00284/REP, pp. 1-59, December 1992.

[156] Lowson, M. V.; Fiddes, S. P.: *Design Prediction Model for Wind Turbine Noise: 1. Basic Aerodynamic and Acoustic Models*. Flow Solutions Report 93/06, W/13/00317/00/00, pp. 1-46, November 1993.

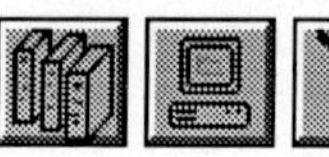

[157] Lowson, M. V.; Fiddes, S. P.; Kloppel, V.; et al.: *Theoretical Studies Undertaken During the Helinoise Programme*. 19th European Rotorcraft Forum, pp. (B4) 1-8, September 1993.

[158] Lowson, M. V.; Lowson, J. V.: *Systematic Comparison of Predictions and Experiment for Wind Turbine Aerodynamic Noise*. Flow Solutions Report 93/03, ETSU W/13/00363/REP, pp. 1-18, April 1993.

[159] Lowson, M. V.: *Theory and Experiment for Wind Turbine Noise*. AIAA Paper 94-0119, 32nd Aerospace Sciences Meeting and Exhibit, pp. 1-9, January 1994.

[160] Lubbers, F.; Pheifer, L. R.: *Final Results on the Research Programme Concerning the Social and Environmental Aspects Related to the Windfarm Project of the Dutch Electricity Generating Board*. Proc. of European Community Wind Energy Conference, Germany, pp. 10-12, March 1993.

[161] Lumely, J. L.; Panofsky, H. A.: *The Structure of Atmospheric Turbulence*. John Wiley, 1964.

[162] MacDonald, H. M.: *A Class of Diffraction Problems*. Proc. of the London Mathematical Society, Vol. 14, No. 2 , pp. 410-427, 1915.

[163] Mantay, W. R.; Campbell, R. L.; Shidler, P. A.: *Full-Scale Testing of an Ogee Tip Rotor*. NASA CP 2502, 1978.

[164] Martinez, R.; Widnall, S. E.; Harris, W. L.: *Predictions of Low-frequency and Impulsive Sound Radiation from Horizontal-axis Turbines*. cited in Hau, E.: Windkraftanlagen, pp. 574, 1988.

[165] Matthies, H.; Garrad, A. D.; Adams, B. M.; et al.: *An Assessment of the Offshore Wind Potential in E.C.* Proc. of the European Wind Energy Association Conference & Exhibiton, Dr. J. L. Tsipouridis (ed.), Thessaloniki, Vol. I, pp. 111-115, October 1994.

[166] Maynard, J. D.; Williams, E. G.; Lee, Y.: *Nearfield Acoustic Holography: I. Theory of Generalized Holography and the Development of NAH*, J. Acoustic Soc. America, Vol. 78, No. 4, pp. 1395-1413, October 1985.

[167] Morino, L.; Tseng, K.: *A General Theory of Unsteady Compressible Potential Flows with Application to Airplanes and Rotors*. Developments in Boundary Element Methods, P. K. Banerjee and L. Morino (ed.), Elsevier Applied Science, Barking, England, UK, Vol. 6, pp. 183-245, 1990.

[168] Morse, P. M.; Feshbach, H.: *Methods of theoretical Physics*. McGraw-Hill, New York, 1953.

[169] Nelson, P. A.; Morfey, C. L.: *Aerodynamic Sound Production in Low Speed Flow Ducts*. Journal of Sound and Vibration, Vol. 79, No. 2, pp. 263-289, May 1981.

[170] Nolle, A.: *Berücksichtigung meteorologischer Einflüsse auf die freie Schallausbreitung in Prognoseverfahren*. Acustica, Vol. 66, pp. 97-103, 1988.

[171] Operti, I.; Preti, D. G.; Ginn, K. B.: *Acoustic Testing in the Automotive Industry using STSF*. B&K Application Note, 1994.

[172] Paterson, R. W.; Vogt, P. G.; Fink, M. R.; Munch, C. L.: *Vortex Noise of Isolated Airfoils*. J. Aircraft vol. 10, pp. 296-302, 1973.

[173] Paterson, R. W.; Amiet, R. K.; Munch, C. L.: *Isolated Airfoil-Tip Vortex Interaction Noise*. AIAA Paper 74-194, 1974.

[174] Petersen, E. L.; Madsen, P. H.: *The Likely State of Wind Technology and its Economics in the Year 2000*. European Directory of Renewable Energy, James & James Science Publishers Ltd., ISBN 1-873936-16-8, ISSN 0266-8041, pp. 147-150, 1993.

[175] Pettersson, F.: *Prediction of Broadband Noise from Wind Turbines with Sawtooth Trailing Edges*. FFA TN 1993-49, pp. 1-62, November 1993.

[176] Pfeiffer, A. E: *Brochure Wind Turbine Noise (Dutch)*. HW 88.1704.1, 1988.

[177] Pinder, J. N.: *Mechanical Noise from Wind Turbines*. Wind Engineering, Vol. 16, No. 3, pp. 158-168, 1992.

[178] Powell, A.: *Mechanisms of Aerodynamic Sound Production.* AGARD Report 466, April 1963.

[179] Powell, A.: *Theory of Vortex Sound.* The Journal of the Acoustical Society of America, Vol. 36, No. 1, pp. 177-195, January 1964.

[180] Preussen Elektra AG: *Information about Electricity.* (in German), 1995.

[181] Rasmussen, K. B.: *Sound Propagation Over Grass Covered Ground.* Journal of Sound and Vibration, Vol. 78, No. 2, pp. 247-255, 1981.

[182] Rehfeldt, K.: *Personal Communication.* Deutsches Wind Energie Institut GmbH, 1996.

[183] Ruscheweyh, H.: *Dynamische Windwirkung an Bauwerken, Band 1: Grundlagen.* Bauverlag GmbH Wiesbaden, pp. 1-96, 1982.

[184] Ruscheweyh, H.: *Dynamische Windwirkung an Bauwerken, Band 2: Praktische Anwendungen.* Bauverlag GmbH Wiesbaden, pp. 1-181, 1982.

[185] Schaefer, H.: *Erntefaktoren von Kraftwerken, Definitionsmöglichkeiten.* Energiewirtschaftliche Tagesfragen, Vol. 38, No. 10, 1988.

[186] Schlinker, R. H.; Amiet, R. K.: *Helicopter Rotor Trailing Edge Noise.* NASA CR-3470, pp. 1-145, November 1981.

[187] Schultz, T. J.: *Social Surveys on Annoyance - A Synthesis.* Journal of the Acoustical Society of America, Vol. 64, pp. 377-405, August 1978.

[188] Seybert, A. F.; Soenarko, B.; Rizzo, F. J.; Shippy, D. J.: *An Advanced Computational Method for Radiation and Scattering of Acoustic Waves in Three Dimensions.* Journal of the Acoustical Society of America, Vol. 77, No. 2, pp. 362-367, February 1985.

[189] Shearin, J. G.; Block, P. J.: *Airframe Noise Measurements on a Transport Model in a Quiet Flow Facility*, AIAA paper 75-509, March 1975.

[190] Shepherd, K. P.; Grosveld, F. W.; Stephens, D. G.: *Evaluation of Human Exposure to the Noise from Large Wind Turbine Generators.* Noise Control Engineering Journal, Vol. 21, No. 1, pp. 30-37, July-August 1983.

[191] Shepherd, K. P.; Hubbard, H. H.: *Measurements and Observations of Noise from a 4.2 Megawatt (WTS-4) Wind Turbine Generator.* NASA CR-166124, pp. 1-33, May 1983.

[192] Shepherd, K. P.: *Detection of Low Frequency Impulsive Noise from Large Wind Turbine Generators.* NASA-CR-172511, N85-16588, pp. 1-22, January 1985.

[193] Shepherd, K. P.; Hubbard, H. H.: *Noise Radiation Characteristics of the Westinghouse WWG-600 (600kW) Wind Turbine Generator.* NASA TM-101576, pp. 1-32, July 1989.

[194] Stearns, S. D.: *Digital Signal Analysis.* Hayden Book Company, Inc., Rochelle Park, New Jersey, 1975.

[195] Stechow, L.; Betke, K.; Schultz-von Glahn, M.; Klug, H.; Kramkowski, T.: *Messung der Schallabstrahlung von Rotorblättern im Windkanal.* Proc. 2. Deutsche Windenergie-Konferenz, Wilhelmshaven, 1994.

[196] Stephens, D. G.: *Guide to the Evaluation of Human Exposure to Noise from Large Wind Turbines.* NASA-TM-83288, N82-24051, pp. 1-71, March 1982.

[197] Stevenson, R.: *Environmental Aspects of Wind Farms.* European Directory of Renewable Energy, James & James Science Publishers Ltd., ISBN 1-873936-16-8, ISSN 0266-8041, pp. 180-183, 1993.

[198] Succi, G. P.: *Design of Quiet Efficient Propellers.* SAE 790584, Business Aircraft Meeting and Exposition, pp. 1-14, April 1979.

[199] Succi, G. P; Munro, D. H.; Zimmer, J. A.: *Experimental Verification of Propeller Noise Prediction.* AIAA Journal, AIAA80-0994R, Vol. 20, No. 11, pp. 1483-1491, November 1982.

[200] Taylor, D.: *Wind Energy.* Renewable Energy, The Open University, T 521, A Resource Pack for Tertiary Education, Section 8 - wind energy, pp. 8(1-40), 1994.

[201] Thompson, R. J.: *Ray Theory for an Inhomogeneous Moving Medium.* The Journal of the Acoustical Society of America, Vol. 51, No. 5 (part 2), pp. 1675-1682, 1971.

[202] Thompson, D. H.: *A Flow Visualisation Study of Tip Vortex Formation.* Aerodynamics note 421 of the department of defence of Australia, 1983.

[203] van der Borg, N.; Andersen, B.; Jakobsen, J.; et al.: *Noise From Wind Turbines.* Final report of JOULEII project JOU2-CT92-0124, Netherlands Energy Research Center report ECN-C--95-036, April 1995.

[204] van Ditshuizen, J. C. A.: *Helicopter Model Noise Testing at DNW - Status and Prospects.* Proc. of the 13th European Rotorcraftform, Arles, France, 8-11 September, 1987.

[205] VDI-Richtlinie 2714: *Schallausbreitung im Freien.* VDI-Richtlinien, pp. 1-18, January 1988.

[206] VDI-Richtlinie 2720, Blatt 1: *Schallschutz durch Abschirmung im Freien.* pp. 1-14, Verein Deutscher Ingenieure, Düsseldorf, February 1991.

[207] Viterna, L. A.: *The NASA LeRC Wind Turbine Sound Prediction Code.* NASA CP-2185, pp. 410-418, February 1981.

[208] Voutsinas, S.: *Development of a Vortex Type Aeroacoustic Model of HAWTs and its Evaluation as a Noise Prediction Tool.* Final Report JOU2-CT92-0148, National Technical University of Athens, University Le Havre, April 1995.

[209] Wagner, S.: *Strömungslehre I, II.* Script of Lecture on Fluid Mechanics, Institut für Aerodynamik und Gasdynamik, Universität Stuttgart, 1995.

[210] Westergaard, C. H.; Grabau, P.: *Requirements to Design Tools for Future Large Wind Turbines with High Performance and Low Aerodynamical Noise.* Proc. 2nd European Computational Fluid Dynamics Conference, ECCOMAS, Stuttgart, 1994.

[211] Willshire, W. L., Jr.: *Assessment of Ground Effects on Aircraft Noise: The T-38 Experiment.* NASA Technical Paper 1747, December 1980.

[212] Wolsink, M.; Sprengers, M.; Keuper, A.; et al.: *Annoyance from Windturbine Noise on Sixteen Sites in Three Countries.* European Community Wind Energy Conference, Lübeck-Travemünde, pp. 273-276, March 1993.

[213] *WWW-Server TU Denmark - Wind Energy.* http://www.afm.dtu.dk/wind/, World Wide Web, 1996.

Annex: Description of Tip Planform Shapes

Within the framework of the JOULE II projects, several different tip planforms have been investigated in order to study the effect of tip shape on the generation of high-frequency noise. This annex contains detailed sketches of all tips which have been used.

Three planforms – a standard, an ogee, and a shark-fin tip – have been mounted on the UNIWEX turbine and measured in the wind tunnel of the University of Oldenburg [131], [24]. A set of six tips have been measured on full scale turbines and in the wind tunnel [131], [120].

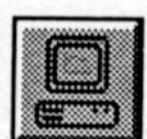

Reference tip

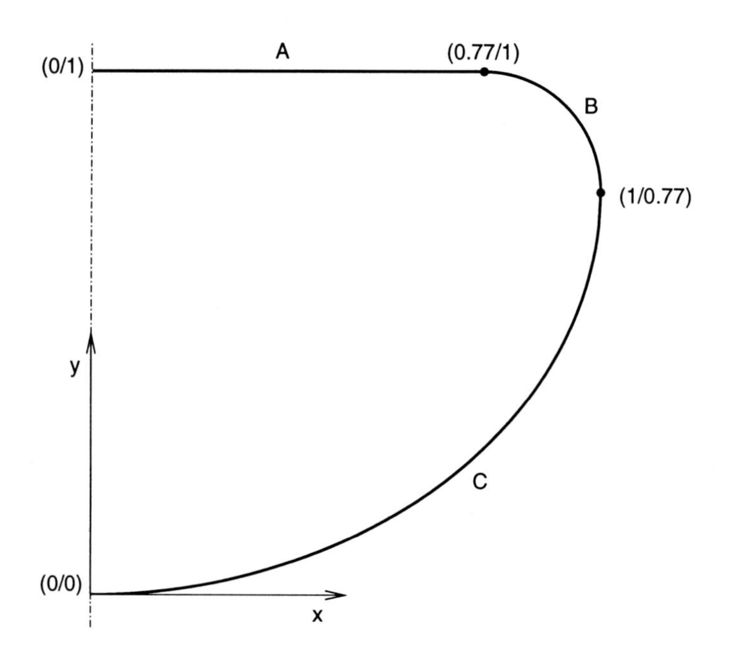

Figure A1: Planform for the reference tip.

The co-ordinates of the tip planform are given for the 3 pieces A, B, and C:

A: $y = 1$ $\quad 0 \le x \le 0.77$ line

B: $y = 0.77 + \sqrt{0.23^2 - (x - 0.77)^2}$ $\quad 0.77 < x \le 1$ circle

C: $y = 0.77 + 0.77\sqrt{1 - x^2}$ $\quad 0 \le x < 1$ ellipse

Ogee tip

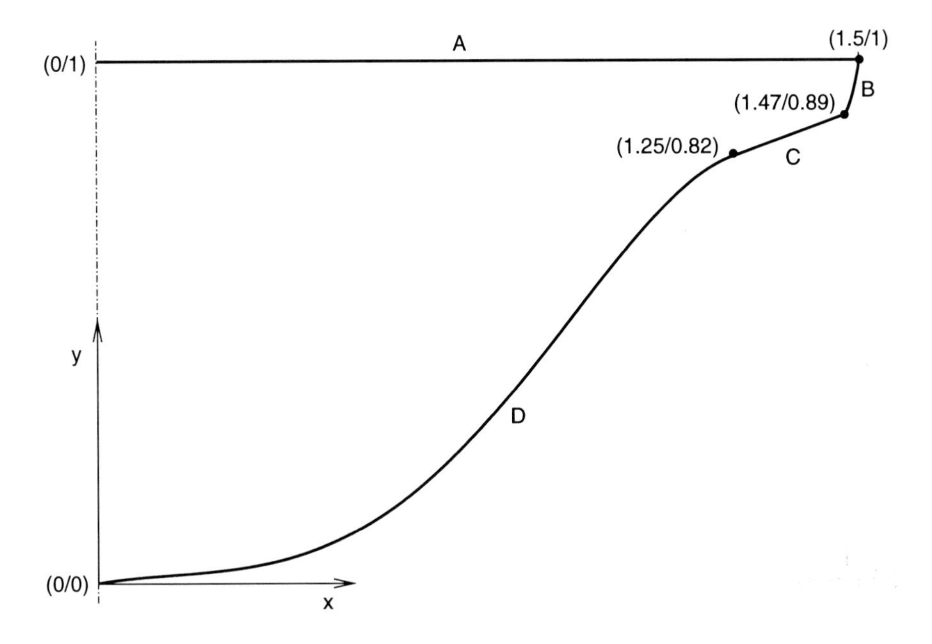

Figure A2: Planform for the ogee tip.

The co-ordinates of the tip planform are given for the 4 pieces A, B, C, and D:

A: $y = 1$ $\quad 0 \le x \le 1.5$ line

B: $y = 75.165x^2 - 219.68x + 161.4$ $\quad 1.47 < x \le 1.5$ parabola

C: $y = 0.36x + 0.3652$ $\quad 1.25 < x \le 1.47$ line

D: $y = -0.48x^5 + 0.14x^4 + 1.37x^3 - 0.6x^2 + 0.16x$ $\quad 0 \le x < 1.25$ polynomial

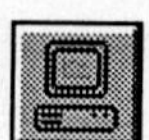

Shark-fin tip

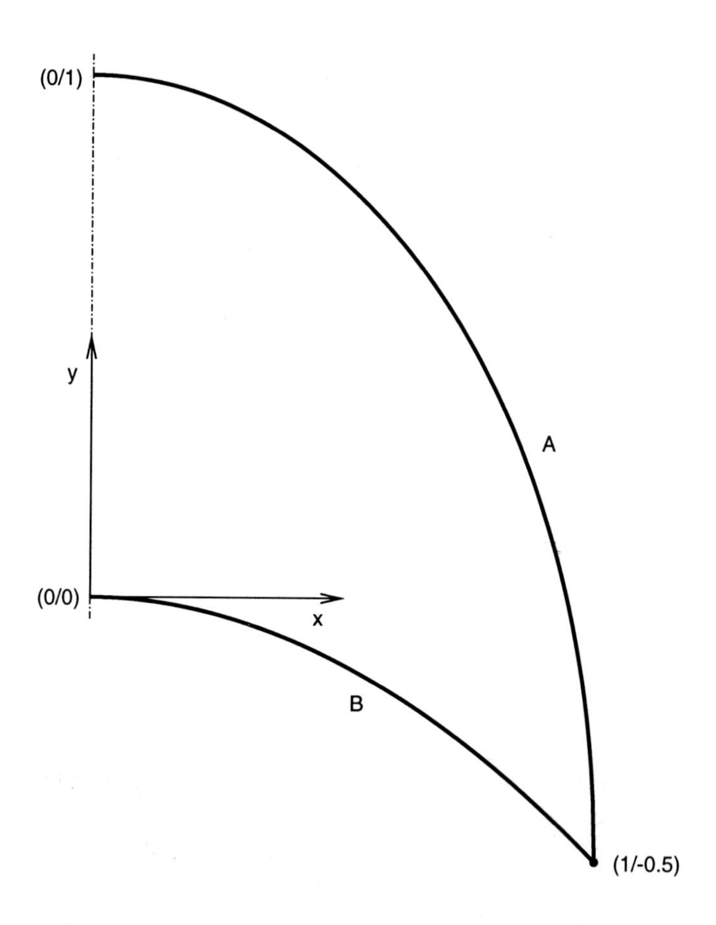

Figure A3: Planform for the shark-fin tip.

The co-ordinates of the tip planform are given for the 2 pieces A and B:

A:	$y = 0.5 + 1.5\sqrt{1 - x^2}$	$0 \le x < 1$	ellipse
B:	$y = -0.5x^2$	$0 \le x \le 1$	parabola

Standard tip

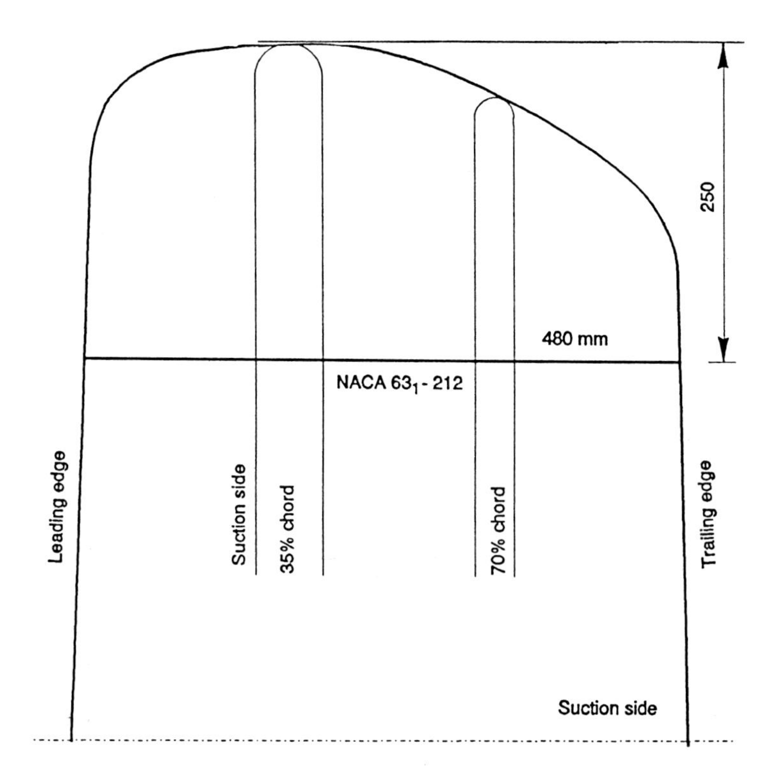

Figure A4: Planform for the standard tip [120].

  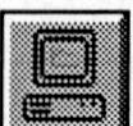 

Tip M

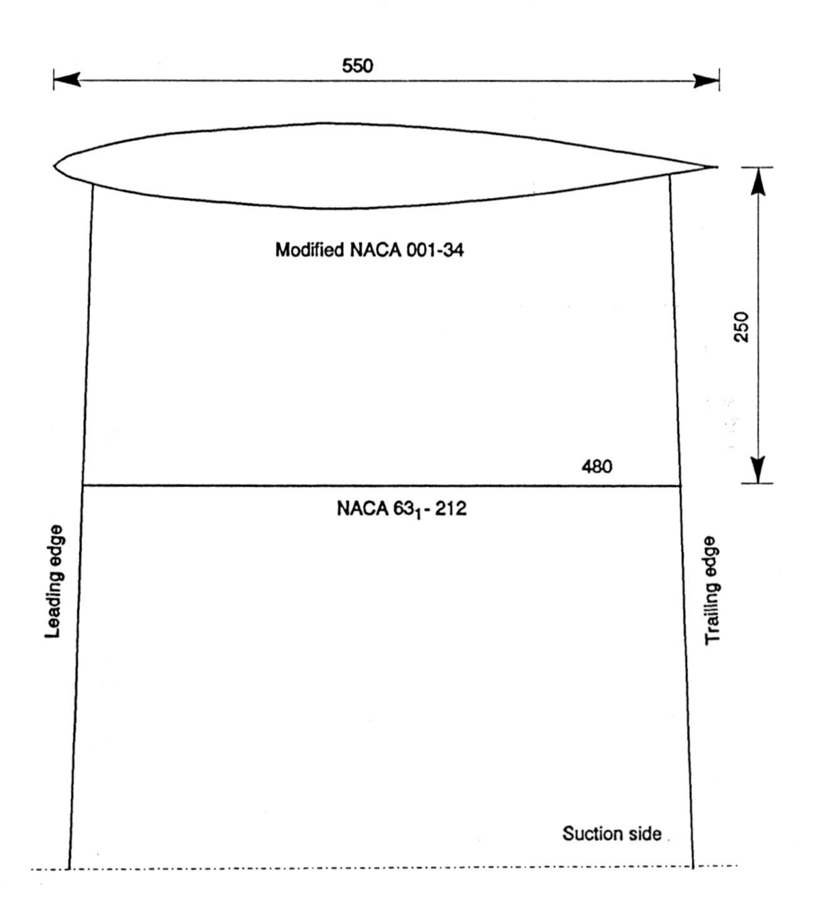

Figure A5: Planform for the tip M [120].

Tip M has a tip torpedo which is supposed to prevent the generation of small scale intense tip vortices by forcing the tip vortices into a larger motion and thereby possibly preventing the return of the vortex on the rotor blade.

Tip J

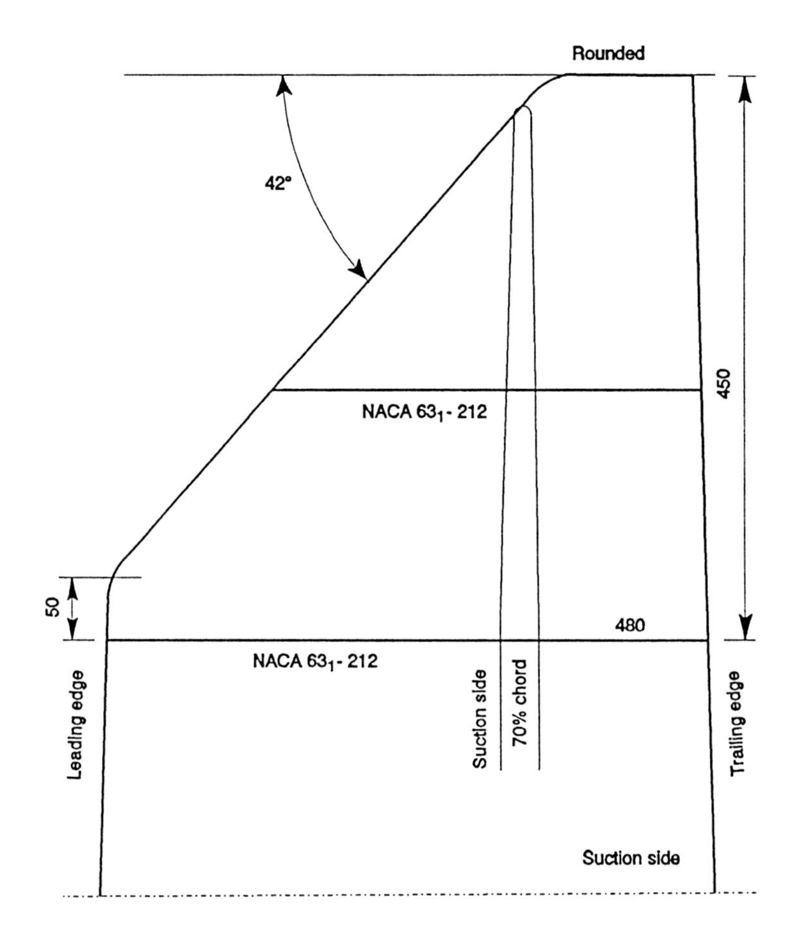

Figure A6: Planform for the tip J [120].

Tip J has a linearly decreased chord from the leading edge and a rounded tip edge. It is aimed at combining the leading edge wake and the tip wake similar to the conditions at a delta wing. The extended trailing edge might give rise to a radial velocity component of the vortex which would cause it to move away outwards from the tip.

Tip I

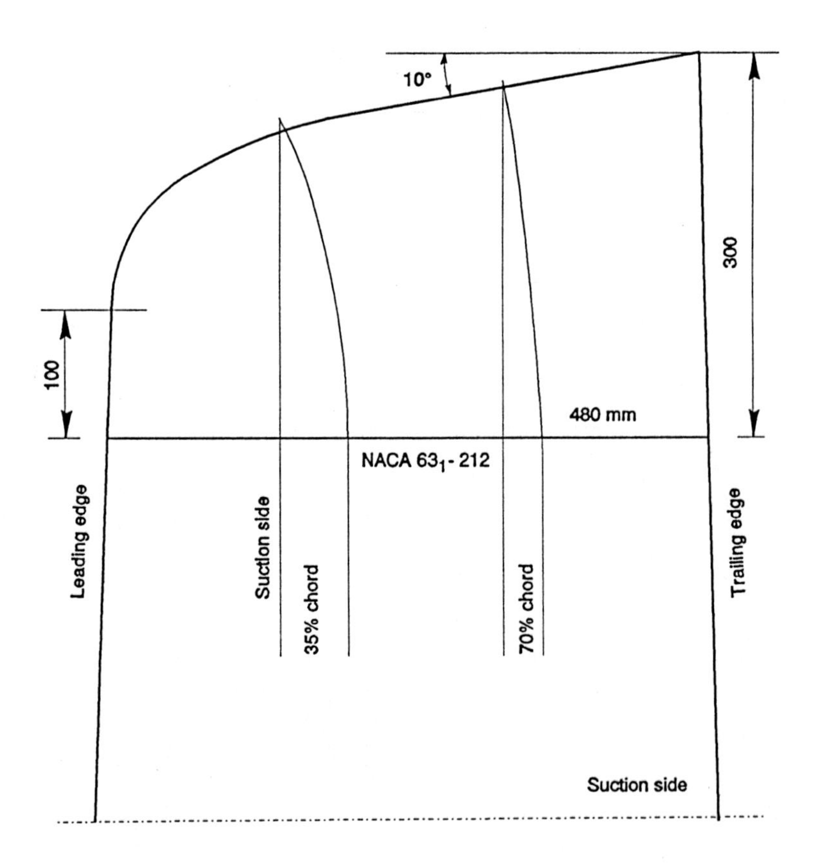

Figure A7: Planform for the tip I [120].

Tip I has an extended trailing edge and a sharp tip edge. Its design is based on the same ideas as Tip J.

Tip R

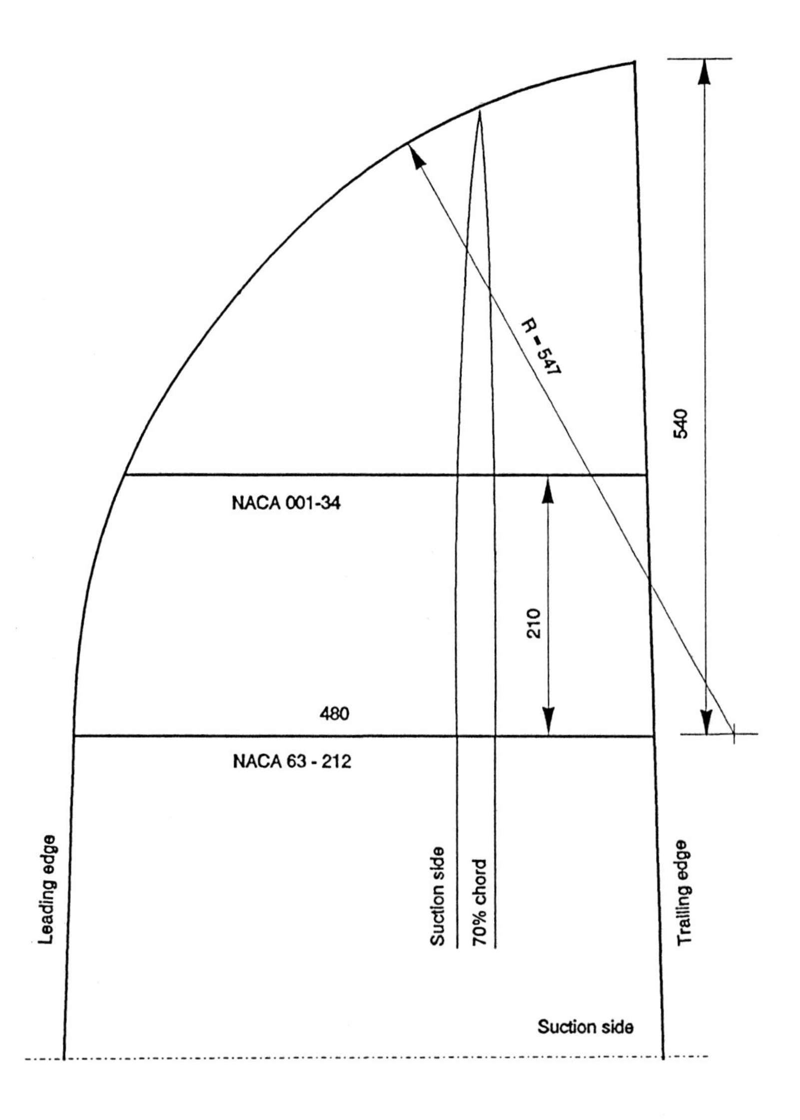

Figure A8: Planform for the tip R [120].

Tip R has a curved leading edge and an extended edge but without the sharp transition for Tip J.

Tip O

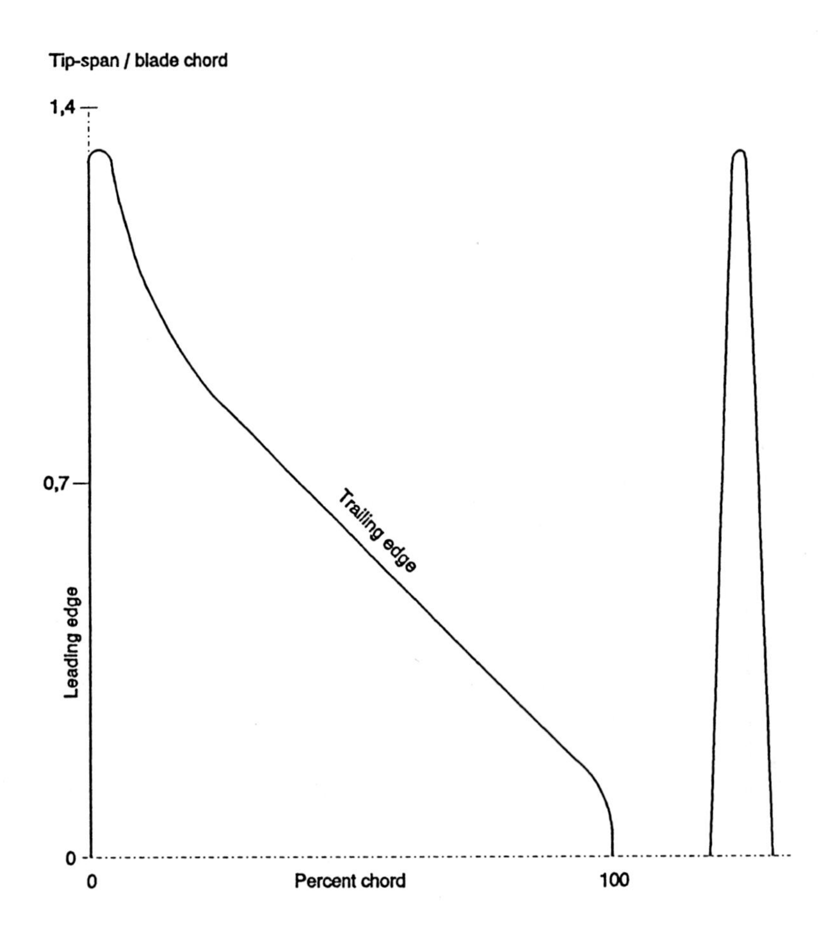

Figure A9: Planform for the tip O [120].

Tip O was inspired from tests with quiet tips for helicopter blades. Due to the extended 'finger' a strong leading edge wake is expected to build up and perhaps being able to disturb the generation of a tip wake. Tip O is similar to the ogee tip (see Section 8.4.2).

European Commission

Final report - "Development of an Aeroacoustic Tool for Noise Prediction of Wind Turbines, Validation on New Full Size Wind Generators"

1996–224p.–15.5 × 23.5 cm

ISBN 3-540-60592-4

Printing: Mercedesdruck, Berlin
Binding: Buchbinderei Lüderitz & Bauer, Berlin